T0314927

Optical Sensing in Power Transformers

Optical Sensing in Power Transformers

Jun Jiang
Nanjing University of Aeronautics and Astronautics, Nanjing, China

Guoming Ma
North China Electric Power University, Beijing, China

Registered Offices
John Wiley & Sons, Inc., 111 River Street, Hoboken, NJ 07030, USA
John Wiley & Sons Ltd, The Atrium, Southern Gate, Chichester, West Sussex, PO19 8SQ, UK

Editorial Office
The Atrium, Southern Gate, Chichester, West Sussex, PO19 8SQ, UK

For details of our global editorial offices, customer services, and more information about Wiley products visit us at www.wiley.com.

Wiley also publishes its books in a variety of electronic formats and by print-on-demand. Some content that appears in standard print versions of this book may not be available in other formats.

Library of Congress Cataloging-in-Publication data applied for

HB : 9781119765288

Cover Design: Wiley
Cover Image: © wolv/Getty Images

Set in 9.5/12.5pt WarnockPro by SPi Global, Chennai, India
Printed and bound by CPI Group (UK) Ltd, Croydon, CR0 4YY

10 9 8 7 6 5 4 3 2 1

Contents

Foreword

I believe that the present book is the first book related to a promising optic-based online monitoring for power transformers in the power grid.

Power transformers are essential equipment for power delivery, which directly influence the security and stability of the large electric network. Therefore, the transformer condition assessment and failure analysis are very important issues for the electric power utilities. There is a very complicated environment in the transformer, which is related to the Multiple Physical Field and its coupling, such as intense electric field, intense magnetic field, heat transfer field, oil flow field, and so on. The current sensing systems are mostly located on the outside of the transformer in order to prevent senses from electromagnetic, chemical, and heating effects, resulting in less physical fields for measuring, low sensitivity for signal acquisition, and low spatial resolution for fault location. It is believed that optical sensing, especially optic fiber sensors, has an advantage of passive, lightweight, small size, EMI immunity, and is able to work in environments with multiple physical fields and chemical corrosion in the transformer. Therefore, I believe that optical sensing technology will be promising technology for a power transformer, not only in an on-line condition monitoring aspect but also in fault protection issues.

This book is aimed to comprehensively present the different cutting-edge optical principles and methodologies adopted for online monitoring of power transformers and covers the basic principle, key points, possible installations, and some results. Although some optical techniques are currently still not mature enough for in-situ applications in power transformers, the book tries to inspire new chances and possibilities. In addition, there are abundant first-hand information, experience, and knowledge on the optical techniques applied in power transformers contributed by Dr. Jiang Jun and Dr. Ma Guoming in this book, which may make it more attractive to the readers.

I really hope this book will be interesting to the audience of scientific researchers, technical R&D staff, manufacturers, frontline engineers, postgraduate students, et al., and we encourage every reader to explore more new possibilities applied in power transformers.

August, 2020

Chengrong Li
Professor
North China Electric Power University

Preface

Transformers are one of the most important pieces of equipment in a power grid. Its health index can significantly impact both the reliability and functionality of the power grid. However, partial in-service transformers worldwide have already reached or exceeded their design life expectancy. Thus, real-time online monitoring and assessment have been prioritized on the agenda among utilities around the globe to allow for a timely maintenance action and to avoid any potential catastrophic failures. Many new detection tools are being investigated continuously by researchers and engineers in the field. In particular, with advances in optical engineering and communications technology, the last few decades have witnessed the emergence and development of a new generation of optical approaches for power apparatus condition monitoring. The inherent advantages of fibre optic sensors include lightweight, compatibility, passivity, low attenuation, low power, immunity to electromagnetic interference (EMI), high sensitivity, wide bandwidth, and environmental ruggedness. These advantages are utilized to compromise for its high cost and unfamiliarity to the consumer. Therefore, they have become commonly used and applied in high voltage applications.

This book presents the concepts and current industry practice of various popular condition monitoring techniques such as temperature measurement, moisture analysis, dissolved gas analysis, partial discharge, winding deformation, vibration analysis, voltage/current measurement, and electric field measurement. The book also provides fundamental knowledge of optical principles and some practical techniques for optical probes design, circuit topology, alternative schemes, and the comprehensive merits and drawbacks. Primarily, the book offers a comprehensive and valuable source of information for researchers, utility engineers, operators, and technicians. It reflects a solid understanding of strategic concepts to maintain assets, optimize planned replacements, and minimize the possibility of catastrophic failures. Finally, it offers advanced material for undergraduate and postgraduate research students, and advanced teaching in the emerging field of advanced sensors and electric engineering.

This book is organized into six chapters.

1. **Chapter 1** gives an overview of power transformers in power grids, their typical structure, as well as an overview of oil-immersed insulation systems. This chapter also provides a foundation for understanding condition monitoring of oil-immersed power transformers to provide a basic outline of traditional techniques and assist the reader in understanding the necessity of novel sensing.

2. **Chapter 2** focuses on temperature detection with optical techniques, alongside heat sourcing and transferring in power transformers. Optical fiber sensors are developed

to detect hot spot temperature (HST) and prevent a "fever" of power transformers, with point measurement, quasi-distributed measurement, and distribution measurement.

3. **Chapter 3** primarily explains moisture measurements by optical sensors, with a comprehensive guideline for three types: FBG, evanescent wave and Fabry–Perot (FP)-based moisture measurements. To build a reasonable and practical online moisture detection, the factors of inferencing the measurement are considered and the pros and cons of each optical technique are weighed respectively.

4. **Chapter 4** presents fundamentals of dissolved gas analysis (DGA) and its requirements for online monitoring. It also shows the categorization of optical schemes for dissolved gas analysis as photoacoustic spectrum (PAS), Fourier transform infrared spectrum (FTIR), tunable diode laser absorption spectrum (TDLAS), laser Raman spectroscopy (LRS), and fiber Bragg grating (FBG). Finally, it provides a comparison between currently used optical fiber techniques.

5. **Chapter 5** concentrates on partial discharge (PD) activities detection with optical techniques on the basis of the PD-induced weak acoustic emission effect. Three main optical techniques, based on different principles, are analyzed for PD detection, namely FBG, FP, and dual-beam interference topology. The sensitivity enhancement, merits, and drawbacks of these techniques are presented as well.

6. **Chapter 6** provides the measurement of mechanical and electrical parameters on the basis of optical solutions, such as winding deformation, high voltage, large current, and contactless electric fields. It also highlights the current condition monitoring limitation with optical techniques and the importance of future research.

The research work presented in this book is supported and funded by the National Natural Science Foundation of China and shows collaborative work and projects from several power utilities in China over a period of about 10 years. The book aims to provide the state-of-the-art knowledge related to optical solutions for condition monitoring and fault diagnosis of power transformers.

Optical fibers may seem irrelevant to power transformers at first glance; however, it is extremely beneficial to combine the interdisciplinary subjects. The authors hope it can be a "must have" reference book and an adequate reference for anyone working with condition monitoring of power transformers. Nevertheless, a continuing effort in the academic research and industrial work is still necessary to emphasize the long-term reliability of the information provided and improve the precision of optical measurements, respectively. Therefore, interests, efforts, opinions, and collaborations from enthusiastic readers are highly appreciated, in order to offer potential feasible solutions and improve the optical presence in the power industry.

Acknowledgments

Many people have supported this work, directly or indirectly, throughout our involvement with the interdisciplinary research and manuscript preparation. We would like to acknowledge some of the key personnel without whose contributions this publication would never have reached this point.

The authors are indebted to Prof. Chengrong Li and Prof. Zhongdong Wang for their generosity to find time in the demanding schedule to offer the Foreword and valuable suggestions for this book.

Mr. Bendong Zhang and Miss Kai Wang at Nanjing University of Aeronautics and Astronautics have contributed directly in preparing the manuscript of this book, especially for their persistent support on the figure drawing and information collection. Their contributions are greatly appreciated. Special thanks go to Dr. Prem Ranjan, Ms. Abir Al abani and Mr. Sanad Elrishe from The University of Manchester for their proofreading of all the chapters.

Jun Jiang would like to express his most sincere thanks to Mrs. Miao Yu, Mr. Wenqing Lu, Mrs. Xiaoyan Wu, and their kids (Leo, Future, and Lucky), who are the strength, support, and inspiration behind each word. He also feels very grateful to the Man_Delta group members for their companionship during the preparation process and COVID-19 quarantine period in Manchester, UK.

In addition, the authors would like to acknowledge some projects for providing the financial and infrastructural support necessary for these research and development works. This work is supported in part by National Natural Science Foundation of China under Grant Nos. 51677070, 51807088, 51977075, in part by Natural Science Foundation of Jiangsu Province under Grant No. BK20170786, in part by Beijing Natural Science Foundation under Grant No. 3182036, in part by the State Key Laboratory of Alternate Electrical Power System with Renewable Energy Sources under Grant No. LAPS19010. The authors also gratefully acknowledge financial support from China Scholarship Council (No. 201906835029), Fok Ying-Tong Education Foundation for Young Teachers in the Higher Education Institutions of China (No. 161053), Young Elite Scientists Sponsorship Program (No. CAST YESS20160004), and 2019 CAST Outstanding International Youths Exchange Program.

Lastly, the authors are grateful for the extensive help provided by Juliet Booker for her dedication, attention to detail, and efforts during the preparation of this book.

About the Authors

Jun Jiang is an Associate Professor with the Jiangsu Key Laboratory of New Energy Generation and Power Conversion, Nanjing University of Aeronautics and Astronautics, China. He was born in Anqing, China, in 1988. He received the BE degree in electrical engineering and automation from China Agricultural University (CAU) in 2011 and PhD degree in high voltage and electrical insulation from North China Electric Power University (NCEPU) in 2016. During 2019–2020, he worked as an Academic Visitor/Honorary Staff in the Department of Electrical and Electronic Engineering, School of Engineering, The University of Manchester, UK.

At present, he is an IEEE Senior Member, Cigre member, and also a representative for Cigre JWG D1/A2.77 (Liquid Tests for Electrical Equipment). He has published more than 60 peer-reviewed papers including more than 40 journal articles. Also, more than 12 patents have been granted. He was granted the *Young Researcher Award* by International Symposium on High Voltage Engineering (ISH) and the *Outstanding Reviewers Award* by IET High Voltage.

His research interests are optical fiber sensing, condition monitoring of power apparatus, and more-electric-aircraft.

Guo-ming Ma is a Professor with the State Key Laboratory of Alternate Electrical Power System with Renewable Energy Sources and School of Electrical and Electronic Engineering, North China Electric Power University. His research interest covers transient measurement, advanced optical sensing, and condition monitoring of power apparatus.

He is an Associate Editor of *High Voltage* published by IET, Senior Member of IEEE, and CSEE.

His research achievement was awarded a 1st Prize of the China Power Technology Progress Award. He was selected for the "Young Elite Scientists Sponsorship Program" by the China Association for Science and Technology, 2017. He was awarded the "Outstanding Young Electrical Engineering Researcher of China" by the Chinese Society for Electrical Engineering, 2017.

As a first-author, co-first author, or corresponding author, he has published more than 90 academic papers. He is co-drafter of two CIGRE technical brochures and one IEEE international standards. Until now, he has been granted 20+ Chinese invention patents.

He received approximately ten fundings from China government-related agencies, and over twenty research contracts from State Grid Cooperation of China and China Southern Power Grid.

Acronyms

A.U.	Arbitrary unit
AC	Alternating current
AE	Acoustic emission
AI	Artificial intelligence
ANSI	American National Standards Institute
AOCT	All-fiber optical current transformer
AOVT	All-fiber type optical voltage transformer
APD	Avalanche photodiode
ASTM	American Society for Testing and Materials
BDV	Breakdown voltage
B-OTDA	Brillouin optical time-domain analysis
B-OTDR	Brillouin optical time-domain reflectometry
BSO	Bismuth silicon oxide
CBM	Condition-based maintenance
CNT	Carbon nanotube
CT	Current transducers
CW	Continuous wave
DAQ	Data acquisition
DC	Direct current
DCM	Differential cross-multiplication
DGA	Dissolved gas analysis
DP	Degree of polymerization
DTM	Duval triangle method
DTS	Distributed temperature sensing
EFBG	Etched FBG
EFPI	Extrinsic Fabry–Perot interferometer
EM	Electromagnetic
EMI	Electromagnetic interference
ESA	Electrical spectrum analyzer
ESDD	Equivalent salt deposit density
FBG	Fiber Bragg grating
FDS	Frequency domain dielectric spectroscopy
FEA	Finite element analysis
FEP	Fluorinated ethylene propylene
FE-SEM	Field emission scanning electron microscope

FET	Field effect transistor
FIB	Focused ion beam
FID	Flame ionization detectors
FLCEAS	Frequency-locking cavity enhanced absorption spectroscopy
FOSs	Fiber optic sensors
FP	Fabry–Perot
FPI	Fabry–Perot interferometer
FRA	Frequency response analysis
FRM	Faraday rotating mirror
FTIR	Fourier transform infrared spectroscopy
FWHM	Full width at half maximum
GC	Gas chromatography
GIS	Gas insulated switchgear
GMM	Gaussian mixture model
HFCT	High frequency current transformer
HITRAN	High-resolution transmission molecular absorption database
HST	Hottest spot temperature
HVDC	High voltage direct current
IEC	International Electrotechnical Commission
IEEE	Institute of Electrical and Electronic Engineers
IFFT	Inverse fast Fourier transform
IFPI	Intrinsic Fabry–Perot interferometer
IR	Infrared spectroscopy
ISAM	Ionic self-assembly monolayer
KDP	Potassium dihydrogen phosphate
KFT	Karl–Fischer titration
LCSET	Lowest cold start energizing temperatures
LD	Laser diode
LED	Light emitting diode
LOD	Limitation-of-detection
LPGs	Long period gratings
LRS	Laser Raman spectroscopy
LTC	Load tap changer
MCU	Micro control unit
MEMS	Micro electro machining system
MFC	Mass flow controller
MIR	Mid-infrared
MMI	Multimode interference
MNF	Micro/nano fiber
M-Z	Mach–Zehnder
NDIR	Non-dispersive infrared
NIR	Near-infrared
NPs	Nanoparticles
OCT	Optical current transducer
OFAF	Oil forced air forced type
OFAN	Oil forced air natural type
OFGs	Optical fiber gratings

OFWF	Oil forced water forced
ONAF	Oil natural air forced type
ONAN	Oil natural air natural type
ONWF	Oil natural water forced
OSP	Optical sensor probe
OVT	Optical voltage transducers
PA	Peak area
PAS	Photoacoustic spectroscopy
Pd	Palladium
PD	Partial discharge
P_D	Photodiode/photodetector
PDC	Polarization depolarization current
PDIV	Inception voltage of partial discharge
PET	Polyethylene glycol ester
PGC	Phase generation carrier
PI	Polyimide
PID	Proportion integration differentiation
PM	Planned maintenance
PMMA	Polymethyl methacrylate
PNNL	Pacific Northwest National Laboratory
PPV	Peak-to-peak voltage
PRPD	Phase-resolved partial discharge
PS-FBG	Phase-shifted FBG
PTFE	Polytetrafluoroethylene
PVA	Polyvinyl acrylate
PZT	Piezoelectric transducer
QCL	Quantum cascade laser
QEO	Quadratic electro-optic
QEPAS	Quartz-enhanced photoacoustic spectroscopy
QTF	Quartz tuning fork
RAM	Residual amplitude modulation
RH	Relative humidity
RI	Refractive index
R-OTDR	Raman optical time-domain reflectometry
RS	Raman spectroscopy
SDM	Space division multiplexing
SERS	Surface-enhanced Raman scattering
SHM	Structural health monitoring
SMF	Single-mode fiber
SNR	Signal-to-noise ratio
SP	Side-polished
SP-FBG	side-polished FBG
TCD	Thermal conductivity detectors
TDCG	Total dissolved combustible gas
TDLAS	Tunable diode laser absorption spectrum
TDM	Time division multiplexing
TEC	Thermoelectric

TF	Time frequency
THC	Total hydrocarbons
TLS	Tunable laser source
TOF	Time-of-flight
TOT	Top oil temperature
UHF	Ultra-high frequency
VT	Voltage transducers
WDM	Wavelength division multiplex
WMS	Wavelength modulation spectroscopy
Φ-OTDR	Phase sensitive optical time-domain reflectometry

List of Figures

List of Tables

1

Power Transformer in a Power Grid

Power transformers are definitely the most important and complicated apparatus in a mature power grid. A transformer is defined as a static or fairly passive piece of electrical equipment by ANSI (American National Standards Institute)/IEEE (Institute of Electrical and Electronic Engineers), with the function of power transferring and voltage conversion based on the principle of electromagnetic induction. The term power transformer is dedicated to large capacity equipment generally with a rating higher than 500 kVA, connecting between the power station and transmission or distribution segment. An illustration of a typical power grid is shown in Figure 1.1. With literally numerous connection points or nodes in an actual operating power network, power transformers are crucial in every interconnection with the need for a transition at different voltage levels or ratings.

Generally speaking, power transformers can fall into three groups according to their power capacity: (i) small scale power transformers: 500–7500 kVA; (ii) medium scale power transformers: 7500 kVA–100 MVA; (iii) large scale power transformers: 100 MVA and above. It is easy to understand that capacity is closely related to working voltage ratings. With increasing demand for electric energy due to regional and global industrialization, large scale power transformers become usual in a power grid.

Even though power transformers are dedicated to a conventional alternating current (AC) power system owing to electromagnetic induction, they also play a dominant role in thriving high voltage direct current (HVDC) grids. HVDC is favored for energy transmission over long distances and asynchronous coupling between AC regional networks, and mainly power electronic circuits are necessary to convert AC to DC (rectifier circuits) or convert DC to AC (inverter circuits). In this way, a power transformer that has one of its windings connected to one of these circuits is specifically called a converter transformer. It serves several major functions: (i) to isolate the AC from DC systems to prevent the DC potential entering the AC system; (ii) to supply controllable AC voltage needed by the converter valves, i.e. to supply AC voltages in two separate circuits with a relative phase shift of 30 electrical degrees for the two series-connected six-pulse bridges of a 12-pulse converter valve to reduce the low-order characteristic harmonics, especially the fifth and seventh harmonics; (iii) to reduce short-circuit currents under faults and control the rate of increase in valve current during commutation.

In summary, power transformers still gain a heart-like position whether in a common AC system or a blooming HVDC grid.

Optical Sensing in Power Transformers, First Edition. Jun Jiang and Guoming Ma.
© 2021 John Wiley & Sons Ltd. Published 2021 by John Wiley & Sons Ltd.

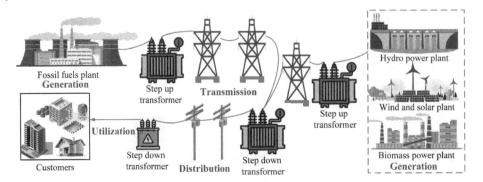

Figure 1.1 Power transformers in a typical power grid/schematic drawing of a power system.

1.1 Typical Structure of a Power Transformer

To begin with, regardless of the sizes or shapes, all power transformers consist of several fundamental components: high voltage coil (also called winding), low voltage coil, and magnetic core. The two coils (also called the primary coil connected to the power source and the secondary coil connected to the load) are around the core to share the magnetic path, with the goal of voltage conversion.

Since the working voltage of a power transformer is always quite high, the insulating safety distances should be guaranteed. It is simple and convenient to install a power transformer outdoor, which has to maintain stable and excellent electrical performances under various working conditions and tests in the natural environment. In this sense, the protective shell is necessary for a power transformer to encapsulate the basic windings, magnetic core, and the related strengthened insulation materials (such as compressed gases, insulation oil, solid epoxy resin) into the steel tank. The shell is always well grounded to avoid the hazard of suspension potential, and the inlet and outlet of the high voltage leads should engage a specific structure with bushings. The inside and outside illustration of a typical power transformer filled with oil insulation can be seen in Figure 1.2. As mentioned above, the working conditions depend on the surroundings, especially the ambient temperature, which has an impact on the volume expansion or contraction; thus an auxiliary oil conservator is placed in the top position.

In addition, frequent variations in load changes the voltage of the system. The tap changes made in the power transformer is mainly done to keep the output voltage within the prescribed limit. Nowadays almost all large power transformers are provided with an on-load tap changer.

As a traditional piece of equipment, it is common sense that a power transformer is gifted with a high efficiency, usually greater than 95% [1], or even 99%. This means that the transformer changes one AC voltage level to another while keeping the input and output powers nearly the same, thus providing a bridge of practically equal power flow from the power source to the loads. Even though a small amount of inefficiency occurs inside the transformer due to leaking magnetic flux outside the core, an eddy current induced by stray magnetic flux in the metallic core structure, Joule heating in copper windings, etc., the amount of heat losses is quite large and leads to a fairly high increase in temperature. Therefore, the cooling concern should be addressed. Different transformer cooling methods, such as that of the oil-immersed transformers, can be

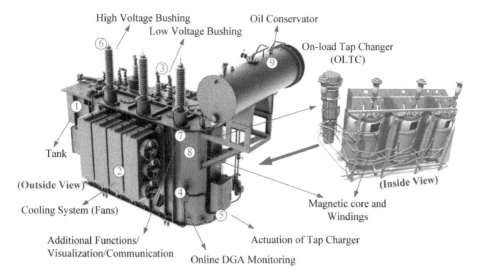

High Voltage Bushing   Oil Conservator

Low Voltage Bushing

On-load Tap Changer (OLTC)

Tank

(Outside View)

Cooling System (Fans)

(Inside View)

Magnetic core and Windings

Additional Functions/ Visualization/Communication

Actuation of Tap Charger

Online DGA Monitoring

Figure 1.2 Typical structure of a large-scale oil-immersed power transformers.

divided into the following types: Oil Natural Air Natural type (ONAN), Oil Natural Air Forced type (ONAF), Oil Forced Air Natural type (OFAN), Oil Forced Air Forced type (OFAF), Oil Natural Water Forced type (ONWF), and Oil Forced Water Forced type (OFWF). The selection is determined by the scale and efficiency of the transformers, the ambient temperature, cooling conditions, etc. Usually, a cooling fan is mounted on the shell to control the temperature increase.

1.2 Insulation Oil in a Power Transformer

Since a power transformer acts as the voltage conversion unit, a high voltage input/output always exists. The insulation system is a critical part in the complex equipment and its construction mainly depends upon its design and application. Usually, a high quality mineral oil acts as the insulation oil that is able to stand up to the weather outside, while transformers intended for indoor use are primarily of the dry type but with also a small of portion of immersed liquid. After the first transformers built in the 1880s, mineral oil was proposed to be used as a transformer cooling and insulating medium [2]. However, under severe conditions, like outdoor use and an extremely high voltage level, transformers are usually immersed in a liquid. Over time the materials have improved dramatically, but the basic concept has changed very little, with insulating oil surrounding a transformer core–coil assembly that enhances the dielectric strength of the winding and prevents oxidation of the core.

Compared with non-lubricated type transformers, oil picks up heat while it is in contact with the windings and carries the heat out to the tank surface by heat transfer or self-convection. Also, the insulation oil enables the transformers to operate at much higher tolerances (e.g. overload or high temperature for a short time) as well as far more reliably, proving that it is the best solution for any situation, especially in the higher voltage levels. It is cost-saving as well since a transformer immersed in oil can

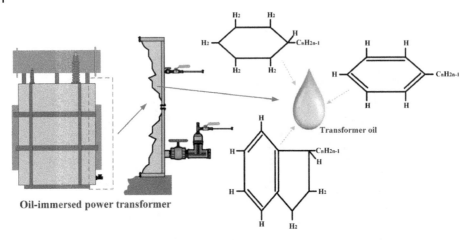

Oil-immersed power transformer

Transformer oil

Figure 1.3 Illustration of typical components of typical transformer oil. Source: Reprinted with permission from J. Jiang et al. [3]. © 2019, Elsevier.

have smaller electrical clearances and smaller conductors for the same voltage and kVA ratings.

There is no doubt that mineral oil is the best component of insulation fluid, owing to its stable properties, and cumulative knowledge and experience about its properties have been found over the decades. In essence, transformer oil is a mixture of a variety of large hydrocarbon molecules, including isoparaffinic, naphthenic, naphthenic-aromatic, and aromatic hydrocarbons, as illustrated in Figure 1.3.

In addition, vegetable oil and compressed gases have been applied in recent years. Given the increasing pressure of economic and ecological awareness, anti-flaming vegetable oil has been researched and considered as an alternative. Especially, the natural ester of vegetable oil has been brought into several large scale power transformers, up to 420 kV for the commercial type (at Bruchsal-Kändelweg substation in Germany by Siemens, https://www.power-technology.com/news/newssiemens-commissions-worlds-first-420kv-vegetable-oil-transformer-4185904/), proving the advantages of much less flammable and completely biodegradable when compared to mineral oil. To date, there are mainly three types of transformer oil used in transformers: mineral oil (paraffin-based transformer oil and naphtha-based transformer oil), synthetic oil (e.g. silicon oil), and vegetable oil (includes natural ester and synthetic ester, e.g. soya, rapeseed, and sunflower oil). Whatever the insulation oil is composed or consists of, generally there are four necessary properties that should be considered to meet the serviceability in power transformers:

1. High dielectric strength.
   There are several indexes to weigh and evaluate the quality of transformer oil: dielectric strength, specific resistance, moisture, and the dielectric dissipation factor. The dielectric strength is directly related to the insulation property of the oil and its value can be experimentally obtained through high voltage breakdown. It is therefore also called the breakdown voltage (BDV) of transformer oil. A recommended device (IEC 60156), mainly a pair of separated metal electrodes with a specific gap immersed in the oil, can give a reading of the BDV. Undoubtedly, a relatively high value of BDV

stands for high dielectric strength, while a low value means unexpected dielectric strength, probably due to some moisture content and conducting substances in the oil. The breakdown field strength of the liquid medium for practical application generally ranges from 20 to 25 kV/mm.

2. Low viscosity and high thermal conductivity.

   Apart from insulation, oil plays a significant role of cooling in a power transformer or other oil-filled power apparatus. In this sense, the fluids in a power transformer are expected to act as an electrically insulating medium and heat transfer agent.

   To ensure the effect of heat exchange, low viscosity and high thermal conductivity are therefore desirable. Literally, general heat transport properties of the oil, especially thermal conductivity, are taken into consideration as important performance criteria in the selection of the transformer fluid according to some specific application. It is a heat transfer coefficient to describe the degree to remove the heat inside, with a unit of W/(m · K). Viscosity is one of factors that influences thermal conductivity of transformer oils, with the unit of Pa · s. In addition, a high boiling point is essential to allow for the continuous liquid state at a relatively high temperature. With the heat transfer fluid, oil-immersed power transformers could easily run in a heat balancing status and even hold a high temperature for extended periods.

3. Resistance to oxidation.

   In practice, oil oxidation inevitably occurs due to oxygen, heat, moisture, and is facilitated by the presence of copper in a power transformer. Especially given the enormous heat and severe electric stress, the oil oxidation process is always on the go during the operation. What is worse, a paper insulation system would be attacked by the by-product of oil oxidation, such as free radicals, acids, water, alcohols, peroxide, aldehydes, ketones, and esters. Accordingly, oxidation inhibitors are added in all transformer oils to some extent. However, the content of the inhibitor substance should be strictly limited to comply with standards like IEC 60296, ASTM D3487, and AS1767. Although it is commonly understood that the oxidation stability of uninhibited oil will decline "gradually" over time, the inhibitors play an important role in preventing the oxidation from declining sharply. Sometimes, re-refined or regenerated transformer oil is brought into the tank to replace the old liquid and to improve the oxidation stability and extend the service time.

4. Low temperature property.

   Despite the typical flash point of the transformer oil being 140 °C or greater due to the heat transfer, the low temperature also influences the property and activity of the oil, known as the pour point. It is the minimum temperature at which oil starts to flow under standard test conditions. The pour point of the transformer oil is an important property, mainly at places where the climate is icy. If the oil temperature falls below the pour point, the transformer oil stops convection flowing and obstructs cooling in the transformer. Typically, the pour point of the power transformer oil is −30 °C or lower, satisfying almost all the chilly conditions with the consideration that there will be some temperature rise during the operation. Transformer oils are graded in line with international standards such as IEC 60296. Within these gradings, there are usually a range of classes, defining transformer oils usage, additives, the lowest cold start energizing temperatures (LCSETs), and more.

Some properties of several transformer oils are shown in Table 1.1 [4].

Table 1.1 Insulating properties of different types of transformer oils.

Fluid type	Flash point, °C	Fire point, °C	Class	Class	Net calorific value, MJ/kg	Ester linkages	Approx. water absorption @ 23 °C
Mineral oil	160–170	170–180	O	1	≥ 42	0	55
Silicon oil	>300	>350	K3	3	≥ 32	0	220
Natural ester	>300	>350	K2	2	≥ 42 and <32	3	1100
Synthetic ester	>250	>300	K3	3	<32	4	2600

Note: O Class ≤ 300 °C, K Class >300 °C according to IEC 61100.
Source: Reprinted with permission from Gnanasekaran and Chavidi [4]. © 2018, Springer Nature Switzerland AG.

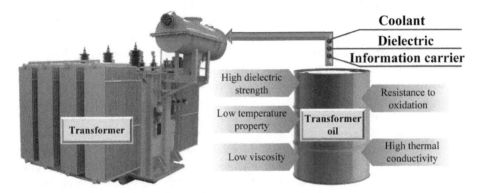

Figure 1.4 Roles and properties of insulation oil in power transformer.

Insulating oil is as important to transformers as blood is to a body. Thus we can get access to the health status of the transformer through necessary oil testing. To summarize, we can find that insulation oil serves multiple roles in the power transformer: dielectric, coolant, and information carrier, as shown in Figure 1.4.

To ensure the availability of the insulation oil, transformer oil needs to be tested to meet the requirements and be up to data standards. Oil testing consists of measuring the BDV and other chemical and physical properties of the oil, either through portable test equipment or in a laboratory. In fact, there are plenty of testing procedures and standards are defined by several international institutes, such as IEC and IEEE, but most of them are set by the ASTM (American Society for Testing and Materials), for example, ASTM D 877 for Dielectric BDV and ASTM D1169 for Specific Resistance.

Since the transformer oil usually lasts for decades stably in ideal state, and proper testing periodically helps to prevent untimely failures and maximize safety or determine whether regeneration or filtration is needed, it saves thousands of dollars in the long term. In many power systems, the key parameters of the transformers such as oil and winding temperatures, voltages, currents, and oil quality as reflected in gas evolution are monitored continuously. This strategy of monitoring can help to prolong the potential operating life of a transformer.

1.3 Condition Monitoring of an Oil-Immersed Power Transformer

As a cornerstone of the power grids, power transformers are one of the most critical and expensive pieces of equipment of a power system, constituting around 60% of substation capital costs, and their proper functionality is vital for the substations and utilities. It is easy to understand that the reliability of the energy system is considerably influenced by its crucial equipment. Therefore, a reliability model of the power transformer is very important in the risk assessment of the engineering systems. As such, maintaining their health and performance levels is critical to everything related to power consumption, from keeping the lights on to industrial production.

Generally, a power transformer is designed and destined for long-term operation, and the normal life expectancy of a power transformer is generally assumed to be about 30 years of service when operated within its rating. However, situations that might involve operation beyond its rating include emergency rerouting of load or through-faults prior to clearing of the fault condition. Therefore, regular and periodical inspection, testing, and maintenance is necessary to carry out in order to find the failures and avoid crash or breakdown cascade events. Corrective maintenance or reactive maintenance often relies on the scheduled or planned outage; thus the maintenance cycle should not be frequent to avoid great operating costs. Moreover, proactive maintenance is a better choice. With the help of emerging sensors and monitoring systems, they can help to decrease the transformer life cycle costs and to increase the high level of availability and reliability. Ultimately, a strategy of condition-based maintenance (CBM) is proposed, which evaluates the premature fault of a sub-health status when certain indicators show signs of decreasing performance or upcoming failure. In a planned maintenance (PM) stage, maintenance work should be performed based upon predefined scheduled intervals. By contrast, CBM is carried out only after observing a decrease in the condition of the equipment on the readings of several parameters. The distinctive difference is the time and decision of maintenance on an as-needed basis, which is timely and cost-effective. As a result, the time to servicing gets shorter and the total cost (including operating cost and maintenance cost) is optimized, as shown in Figure 1.5. CBM is extremely suitable to be applied to the most valuable utilities in a power grid, the power transformers.

Hitherto, a large amount of condition-related sensors and data is necessary to achieve CBM, probably including non-invasive measurements, visual inspection, performance data, and scheduled tests. Therefore, an online monitoring system is becoming popular and lays the data foundation to a preventive maintenance strategy. As to the oil-immersed power transformers, we pay attention to several common parameters necessary to get access to the status of power transformers: temperature, moisture, dissolved gases, vibration, and partial discharges (PDs).

1.3.1 Temperature

An inordinate temperature rise in a power transformer due to load current is known to be the most important factor in causing rapid degradation of its insulation and decides

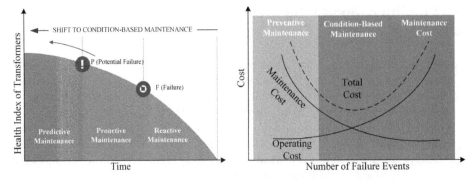

Figure 1.5 Time and cost analysis of condition-based maintenance.

the optimum load carrying ability or the load ability for a transformer. The thermal performance of the power transformer is a crucial indicator when we assess its load ability and usable life. Thermal aging involves the progress of chemical and physical changes as a consequence of chemical degradation reactions, polymerization, depolymerization, diffusions, etc. It also involves the thermomechanical effects caused by the forces due to thermal expansion and/or contraction. The rate of thermal aging and the aging caused by thermomechanical effects, as far as chemical reactions are concerned, are very much influenced by the operating temperature.

Literally, the influence of temperature on the rate at which paper degrades is most commonly taken as Fabre and Pichon's ten-degree rule, where the life halves for every $10\,°C$ increase in temperature. Thus, the accurate measurement of the winding's hottest temperature is critical for calculating the insulation's rate of aging.

For the top oil temperature (TOT) and the hottest spot temperature (HST), being natural outcomes of this process, an accurate estimation of these parameters is of particular importance. The accuracy of the prediction is not always as good as desired. An unacceptable temperature rise may occur due to several fault conditions other than overloading, and hence the need for an online monitoring of the transformer becomes more prominent. The on-line monitoring of a transformer's condition or rather of the condition of its insulation involves recording the voltage and current as well as monitoring the TOT and, on this basis, estimation of the windings' hot spot temperature to determine the paper insulation deterioration rate. In addition, some operational problems such as the malfunction of pumps or fans or pollution of coolers can be detected with the temperature information. In particular, the temperatures of the main tank, load tap changer (LTC) tank and windings are important parameters for online monitoring.

Apart from conventional temperature measurements, a dissolved gases analysis can be another indicator to reflect the occurrence of a thermal fault inside the oil tank. For example, acetylene is only produced at very high temperatures that occur in the presence of an arc.

1.3.2 Moisture

Undoubtedly, moisture or water content in transformer oil is highly undesirable as it affects the dielectric properties of the oil adversely. In addition, the paper is highly hygroscopic and absorbs water from oil. Then the oil/paper insulation system becomes wet, the insulation performance is reduced, and the aging process of the insulation materials is accelerated. The increase of moisture in transformer oil will lead to a significant decrease in the insulation ability of the transformer oil, which will threaten the safe operation of the transformer. Therefore, the detection of moisture in transformer oil is still gaining importance with regard to the state-based maintenance of oil-immersed transformers. Nowadays, the detection of moisture in transformer oil mainly includes two categories [5–7], i.e. offline detection and online monitoring. Since the moisture is widely distributed in ambient surroundings, the correctness and non-contamination of the moisture measurement is hard to guarantee for offline operation. Although the moisture content of an in-service transformer is not expected to vary rapidly [8], the continuous online measurement helps to account for diffusion time and to improve significantly the moisture content calculation.

1.3.3 Dissolved Gases Analysis

Oil constitutes a major component of transformers, so it is easy to get an insight into the inside insulation condition through the oil index. No matter which kind of insulation oil is used, mineral or synthetic or vegetable oils, mainly composed of hydrocarbons, it will contain carbon (C) and hydrogen (H) atoms linked by C—C and C—H, and C—O chemical bonds. The chemical bonds are easily broken when thermal or electrical faults are stressed over time. Active H atoms and hydrocarbon fragments are rearranged to form small molecular hydrocarbon gases, typically methane (CH_4), ethyne (C_2H_2), ethene (C_2H_4), ethane (C_2H_6), carbon monoxide (CO), and carbon dioxide (CO_2). Since all mentioned gases must have an unpaired electron when the breakdown occurs, this means that these gases are in fact the result of a secondary chemical reaction. Since these gases can reveal the faults of a transformer, they are seen as the symbol product and referred to as key gases, also known as "fault gases." Over the past decades, with the help of the informative indicator to the oil state, dissolved gas analysis (DGA) monitoring has become a very useful diagnostic tool and is being universally applied by the utilities or manufacturers for condition assessment of a power transformer and even load tap-changers or bulk oil breakers [9, 10]. Gases are produced by oxidation, vaporization, insulation decomposition, oil breakdown, and electrolytic action. Typical gas combinations produced at various fault types can be found in Table 1.2.

In the early stages, the gas chromatography (GC) method is mostly used as a quantitative measurement in the laboratory and default offline routine tests. DGA usually consists of sampling the oil and sending the sample to a laboratory for analysis. Also, mobile DGA units can be transported and used in the field. To date, it is still being used and provides standard results for the concentrations of dissolved gases. However, the oil cannot keep in a stable form for a long time and is vulnerable to transportation and exposure to air. As a better choice, online monitoring of DGA became popular and works as an essential integral component for oil-immersed power transformers, especially in the era of CBM for Smart Grids. The detector unit is directly connected with the to-be-measured equipment, and the nearly-real-time data of fault gases is sent to the detection center through cable, fiber, or network, beneficial to making decisions and timely feedback. Thus, there is no doubt that the precise detection of dissolved gases is the cornerstone of reliable judgment on the potential defects in power transformer oil.

Table 1.2 Gases produced at different fault types.

Fault type	Key gases	Minor gases
Overheated oil	CH_4, C_2H_4	H_2, C_2H_6
Overheated oil and cellulose	CH_4, C_2H_4, CO, CO_2	H_2, C_2H_6
Partial discharge in oil-paper system	H_2, CH_4, CO	H_2, C_2H_6, CO_2
Spark discharge in oil	H_2, C_2H_2	—
Arcing in oil	H_2, C_2H_2	CH_4, C_2H_4
Arcing in oil and cellulose	$H2, C_2H_2, CO, CO_2$	CH_4, C_2H_4

At the same time, various techniques for detection of dissolved gases in transformer oil have been proposed and implemented.

The insulating oil is in contact with the internal components in the power transformers. The symbol gases dissolved in the oil are formed by normal and abnormal events within the transformer. Therefore, plenty of diagnostic information can be collected by analyzing the concentration, types, proportions, and rate of production of dissolved gases. Several empirical DGA interpretation methods have been utilized in practice. These interpretation methods are based on key gases, key ratios, and graphical representations analysis [11–13], such as IEC, Rogers, Duval' Triangle, and the Key Gas method.

1.3.4 Partial Discharge

Partial discharge (PD) is another key indicator for the insulation condition, which is widely used for preventing the in-service failures of the power transformer. Although partial discharge is defined as a localized dielectric breakdown of a small portion of the electrical insulation under high electric field stress, sustained PD activities might cause irreversible damage and thorough breakdown of the insulation system. Therefore, PD measurement, including PD magnitude (in pC), the number of PD pulses per power cycle, the position of phrase, the possible PD type, and the PD location [14–16], are the prime concern for health monitoring and diagnostic of the transformer.

Detecting PDs is the first stage in a condition-based monitoring system. In terms of PD measurement, the main techniques contain pulse current detection, acoustic emission (AE) detection, a high frequency current transformer (HFCT), and ultra-high frequency (UHF) detection. Pulse current detection is suitable for calibration in the laboratory since it is susceptible to electromagnetic interference (EMI), and the remainder have been widely used in field applications. It is noteworthy that AE-based PD detection has shown good performance to resist EMI and offers a great advantage, having the ability to locate the PD activity position in power transformers.

Moreover, the analysis of PD is quite important when evaluating the insulation defects since the PD activities are a random phenomenon to a certain degree. To obtain the fingerprints of PD, statistical phase-resolved partial discharge (PRPD) is commonly utilized to compare the outline shape, PD pulses and their phrases, etc. Additionally, time-resolved tools help to obtain the valid information based on the extraction of the time and spectral characteristics of the signal, such as mapping of time-frequency (TF). In order to identify the PD type in power transformers, typical feature representation with artificial defects is integrated into the algorithms.

PD detection is a powerful approach used to diagnose the insulation defects in power transformers in the early stages and probably determines the source locations acoustically. On the basis of continuous monitoring of PD, it allows access to the health status of the insulation, thus helping to estimate its lifespan.

1.3.5 Combined Online Monitoring

The available transformer online condition monitoring systems play a useful role in preventive maintenance and life extension [17–19]. Since different parameters focus on the specific fault types in transformers, it is necessary to combine different parameters to get a comprehensive detection of transformers. There are two forms of combination: different techniques can be used together to get accurate results, especially for PD measurement, since the randomness of discharge activities and several related parameters can be combined to diagnose a specific fault. Like water intrusion in oil-paper insulation, the moisture content would rise and concentration of gases would appear after a certain period. Sometimes, the temperature would be affected as well.

In another instance, DGA is also a possible approach to PD activities based on the chemical reaction induced by discharge energy, and the reading of DGA indicates the presence of low-energy PD. However, DGA cannot provide the information of the geometric position of the PD source. To monitor and localize the PD activities in the field, several different types of PD sensors, such as AE, UHF, and HFCT, are jointly employed to ensure an accurate capture.

One should be cautious when determining the status or maintenance of power transformers due to the crucial impact. Combined and integrated online monitoring systems for transformers offer a comprehensive and effective solution with regard to the complex structure of power transformers.

1.4 Conclusion

Power transformers are important and expensive components in the electric power system. However, transformers can be affected by aging due to thermal, mechanical, electrical, and ambient stress. These effects can gradually degrade the insulation system, which can lead to weak points breaking down or crashing in the insulation system. The deregulation of electric power requires a reduction of the service and maintenance cost of the power utilities. Online monitoring and diagnosis are extremely important in order to provide accessible information of the health status on the to-be-determined power transformers in service. For this reason, it is crucial to know the keys for condition monitoring and diagnosis of power transformers in order to achieve a correct course of action (run, repair, or replace). Monitoring systems can help to decrease the transformer life cycle cost and to increase the high level of availability and reliability.

In the large-scale oil-immersed power transformers, the oil plays an important role as dielectric, coolant, and information carrier. Typically, the parameters of temperature, moisture, DGA, and partial discharge are supposed to be detected during the operation process. The accurate measurement and sensing techniques for the state-related parameters are exactly the cornerstone of online monitoring.

Especially in the past few years, significant progress has been made in the field of optical techniques. In this regard, it is believed that the optical sensing techniques introduce not only some chances but also challenges toward new implementation of the conventional power transformer condition monitoring. Through innovation, measurement of parameters, and especially for the need of online monitoring, these will minimize the harm to high-voltage equipment, promote the sensitivity, and achieve stable development. The impetus is to present the latest combination between the optical techniques and high-voltage equipment testing, making it the foundation for further applications, prototypes, and manufacturing.

The following section of this book attempts to comprehensively present the different cutting-edge optical principles and methodologies adopted for online monitoring of power transformers. This interdisciplinary combination and innovation is currently being investigated in depth and will be widely accepted as a new approach to an insulation system for oil-immersed transformers in the coming future.

References

1 Georgilakis, P.S. (2009). *Spotlight on Modern Transformer Design*. Springer Science & Business Media.

2 Harlow, J.H. (2012). *Electric Power Transformer Engineering*, 3e. CRC Press.

3 Jiang, J., Wang, Z., Ma, G. et al. (2019). Direct detection of acetylene dissolved in transformer oil using spectral absorption. *Optik* 176: 214–220.

4 Gnanasekaran, D. and Chavidi, V.P. (2018). *Vegetable Oil Based Bio-Lubricants and Transformer Fluids*. Springer.

5 Alwis, L., Sun, T., and Grattan, K.T.V. (2013). Optical fibre-based sensor technology for humidity and moisture measurement: review of recent progress. *Measurement* 46 (10): 4052–4074.

6 Laskar, S. and Bordoloi, S. (2013). Monitoring of moisture in transformer oil using optical fiber as sensor. *Journal of Photonics* 2013: 528478.

7 A. Swanson, S. Janssens, D. Bogunovic, et al., "Real time monitoring of moisture content in transformer oil," in *Electricity Engineers Conference*, 2018.

8 C. A2.30, "Moisture equilibrium and moisture migration within transformer insulation systems," WG A2.30, 2008.

9 IEEE Std C57.104TM, *IEEE Guide for the Interpretation of Gases Generated in Mineral Oil-Immersed Transformers*, 2019.

10 Sun, C., Ohodnicki, P.R., and Stewart, E.M. (2017). Chemical sensing strategies for real-time monitoring of transformer oil: a review. *IEEE Sensors Journal* 17 (18): 5786–5806.

11 Bustamante, S., Manana, M., Arroyo, A. et al. (2019). Dissolved gas analysis equipment for online monitoring of transformer oil: a review. *Sensors* 19 (19): 4057.

12 Sun, H.-C., Huang, Y.-C., and Huang, C.-M. (2012). A review of dissolved gas analysis in power transformers. *Energy Procedia* 14: 1220–1225.

13 Tang, X., Wang, W., Zhang, X. et al. (2018). On-line analysis of oil-dissolved gas in power transformers using Fourier transform infrared spectrometry. *Energies* 11 (11): 3192.

14 Kanakambaran, S., Sarathi, R., and Srinivasan, B. (2017). Identification and localization of partial discharge in transformer insulation adopting cross recurrence plot analysis of acoustic signals detected using fiber Bragg gratings. *IEEE Transactions on Dielectrics and Electrical Insulation* 24 (3): 1773–1780.

15 Yaacob, M.M., Alsaedi, M.A., Rashed, J.R. et al. (2014). Review on partial discharge detection techniques related to high voltage power equipment using different sensors. *Photonic Sensors* 4 (4): 325–337.

16 Zhou, H.-y., Ma, G., Wang, Y. et al. Optical sensing in condition monitoring of gas insulated apparatus: a review. *High Voltage* 4 (4): 259–270.

17 Fofana, I. and Hadjadj, Y. (2018). *Power Transformer Diagnostics, Monitoring and Design Features*. Multidisciplinary Digital Publishing Institute.

18 Li, C., Ma, G., Qi, B. et al. (2013). Condition monitoring and diagnosis of high-voltage equipment in China-recent progress. *IEEE Electrical Insulation Magazine* 29 (5): 71–78.

19 Li, S. and Li, J. (2017). Condition monitoring and diagnosis of power equipment: review and prospective. *High Voltage* 2 (2): 82–91.

2

Temperature Detection with Optical Methods

Temperature is a great concern in power transformers, due to the tremendous impact of temperature on thermal degradation of the insulation and life expectancy. According to a relatively simple model for thermal aging, known as the Motsinger equation, high temperatures can result in accelerated aging behavior above a critical temperature of 98 °C [1]. Especially, hot-spot temperature (HST), the highest temperature on the oil or winding, is the most important parameter of the transformer's life estimation.

Therefore, continually monitoring temperature is one of the most important procedures to ensure the functionality, reliability, and operational readiness of a transformer. The most convenient technique to monitor transformers temperature is through thermocouple probes like Pt100, which provides oil temperature information for typical positions, at the top and bottom of the cooling unit. However, these sensors are not reliable owing to the uncertainty induced by electromagnetic interference and the distant installation from the hottest winding part. With consideration for design particularities, the need for direct temperature measurements of small, medium, and large power and distribution transformers is much needed.

Fiber optic sensors also provide an excellent and increasingly cost-effective way to collect information on the health of the apparatus through direct temperature measurement of the transformer windings, which is also recommended by the International Electrotechnical Commission (IEC). The closer the temperature measurement is carried out at the hot spot, the more accurate is the approximation of the measured temperature. For reasons of electromagnetic interference and intrinsic safety, optical fiber sensors are a better choice for a power transformer to achieve adequate temperature monitoring [2, 3]. An accurate temperature measurement allows the utilities to operate transformers at their peak capacity and optimize their performance for short term overloading without compromising their life span. Accurate temperature information also gives an increased knowledge of the operational condition assessment, load planning, asset management, and end-of-life determination.

2.1 Thermal Analysis in a Power Transformer

2.1.1 Heat Source in a Power Transformer

According to the description in ANSI/IEEE C57.12.80, continuous rating of a transformer is "the maximum constant load that can be carried continuously without

Optical Sensing in Power Transformers, First Edition. Jun Jiang and Guoming Ma.
© 2021 John Wiley & Sons Ltd. Published 2021 by John Wiley & Sons Ltd.

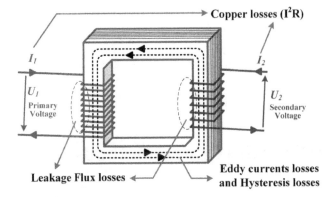

Figure 2.1 Typical losses in an AC power transformer.

exceeding established temperature-rise limitations under prescribed conditions." Various arrays of temperature parameters should be included as temperature limitations, such as core lamination temperature, top oil temperature, ambient temperature, the average winding temperature rise, and the maximum winding temperature rise. It is also expected that the transformer should operate closer to its maximum nameplate capacity with less redundancy in an economic way, but it is inclined to bring in more frequent stress overloading with a direct impact on HST. The maximum temperature or so-called HST is of great importance. Usually, the HST in a power transformer resides in the windings. Therefore, it is often related to the maximum winding temperature rise.

Considering undesirable heat, transformers do not have any moving parts, but they do suffer heat loss. The internal heat of a transformer mainly comes from losses, which comprise the no-load (i.e. iron loss) and load (i.e. copper loss) losses, as shown in Figure 2.1. The no-load loss includes leakage flux losses and core losses present in the iron core in the form of hysteresis and eddy current losses, and is a significant factor of the temperature increase in the magnetic core. With regard to the copper loss, the currents circulating around the transformer's copper windings generate a huge electrical power loss as heat. It is the load loss that affects the temperature rise of the winding the most and the larger capacity of the power transformer brings in more losses and a resultant heat source. As a result, effective cooling measures, apart from natural air cooling, cooling fans, and external oil circulation, should be implemented to remove the dissipated heat of large-scale power transformers.

2.1.2 Heat Transfer in a Power Transformer

There are three heat transfer types in the transformer: conduction, convection, and radiation. In oil-immersed transformers, convection and conduction play a very important role, while the role of heat radiation is relatively small. In transformers with strong oil circulation, the effect of heat radiation is negligible, as shown in Figure 2.2. In natural oil circulation transformers, convection is the main contributor to heat transfer followed by conduction and radiation.

1. Heat conduction
   In oil-immersed transformers, the heat generated by the winding flows into the cooling medium through the insulating coating of the winding wire. In large scale

Figure 2.2 Thermal equilibrium in a power transformer.

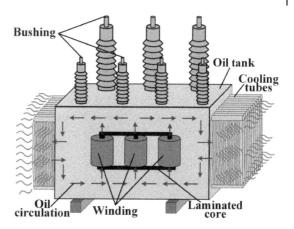

transformers, at least one side of the conductor is in direct contact with the cooling medium, and the heat flows out through a very thin insulating layer. For small scale transformers, the heat usually flows through several insulating layers to reach the cooling medium.

The thermal conductivity of a transformer with oil-paper insulation varies with temperature. In general, reasonably fixed values can be used for engineering calculations. In order to reduce the thermal resistance of the insulating layer, insulating materials with a relatively thin thickness and a relatively large thermal conductivity are preferred in the design of the transformer winding.

A temperature difference between the solid and the fluid drives a heat exchange in the thin layer on the surface between the solid and liquid. The heat exchange within the high/low voltage windings in the transformer versus the oil, the core versus the oil, the oil versus the tank shell, and the tank shell versus the outside air are carried out by thermal conduction.

2. Thermal convection

   The insulating oil has an important property as its volume changes with temperature and pressure. Expansion of the volume with temperature forces the oil to flow in order to achieve the effect of convection cooling. When a heated surface is immersed in a liquid, heat transfers from the solid surface to the cooling medium. In a transformer with natural oil circulation, as the temperature of the fluid increases and the volume expands, the density of the fluid decreases accordingly. Under the action of gravity, the liquid with a small density causes the temperature to rise, and the heat is brought to the surrounding air through the tank wall and the radiator during the rising process. The cool transformer oil with high density then drops down to replace the hot oil. In this way, the transformer oil forms a circulating flow through the process of oil heating up, cold oil supplementation, and hot oil heat dissipating. For a strong oil circulation transformer, the forced flow of oil needs to be much stronger than the buoyancy of the oil caused by the change in density in order to accelerate the oil flow.

3. Heat radiation

   As long as the temperature of any object is higher than the surrounding environment, it will transfer heat to the surroundings through thermal radiation. Although there is a temperature difference in the transformer parts where a heat exchange occurs, thermal radiation is still relatively small compared to conduction and convection.

To enhance the synthetic effect of cooling and ensure a long-term operation, air, water, or oil forced devices are sometimes combined to improve heat conduction and thermal convection.

In summary, when the oil-immersed transformer is in operation, the heat generated by the windings and the iron core causes the transformer oil temperature to continuously rise while the oil density decreases. The oil circulated in the transformer when the hotter oil flows toward the top of the transformer is cooled down by the tube and tank walls while the cool oil flows in a downward direction. Consequently, heat is continuously transferred to the surrounding air through the tubes and the oil tank surface. The heat dissipated through the external surface achieves a state of thermal equilibrium in the temperature field.

Since the temperature distribution is not uniform within a transformer, the insulation paper will ordinarily undergo the greatest deterioration when exposed to high temperatures. Should the hot spot exceed given limits, the rate of deterioration of the solid and liquid insulation system in the transformer will accelerate rapidly.

It is of merit to evaluate the temperature, especially the hottest spot temperature in a complex power transformer, using a finite element analysis (FEA) and its coupling calculation. A finite element multi-physics coupling calculation provides two forms of coupling, viz. one-way and interactive. The one-way type requires one type of calculation to be initially performed and then the result is substituted for another type. A typical process is illustrated in Figure 2.3. On the other hand, interaction requires two different types of calculations to run simultaneously where both results are then substituted to carry out multiple interactions. Even though the interaction process is considered to be more detailed and precise, it is inefficient due to the lengthy calculations.

Since the copper loss mainly occurs in the winding, the IEC recommends a direct winding measurement for the design, test, loading, and maintenance. Therefore, direct winding temperature sensing is extremely essential to assess and optimize power transformers for short term overloading and for failure detection. Due to the enormous merit for intrinsic safety and immunity to electromagnetic interference, optical techniques offer an excellent solution to the direct temperature measurement for complex designed power transformers of all types. Point sensing fiber optic sensors are the most commonly used instruments for hot spot monitoring. Furthermore, the distribution measurement of temperature makes it possible to map the temperature inside the transformer in order to locate possible overheat areas. Thus, the benefits gained from direct temperature monitoring of optical solutions are easily paid back in a power transformer to keep it in a normal state and extend its life span.

2.2 Fluorescence-Based Temperature Detection

2.2.1 Detection Principle

The principle of a fluorescent fiber optic temperature sensor is primarily based on photoluminescence. When the fluorescent material is irradiated by a certain wavelength of incident light, it absorbs the light energy and enters the excited state from the ground

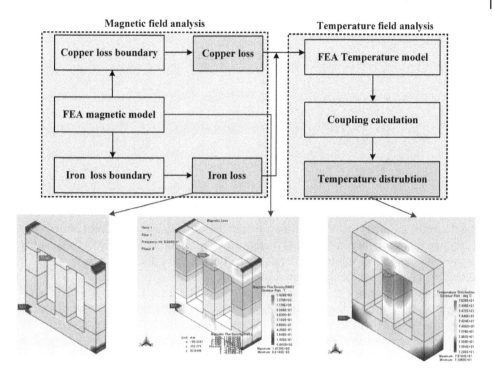

Figure 2.3 Process to calculate the temperature distribution using a coupling FEA model.

state, and immediately de-excites and emits an exit light longer than the wavelength of the incident light. According to Planck's theorem, when the luminescent material receives any form of incident light energy, the electrons in the luminescent material undergo an energy level transition phenomenon, and the process of the energy level transition is accompanied by the exit light of wavelength λ:

$$E_1 - E_2 = hv = \frac{hc}{\lambda}, \tag{2.1}$$

where E_1 and E_2 are the energies of electrons at high and low energy levels, h is Planck's constant, v is the frequency of the outgoing light, c is the speed of light in vacuum, and λ is the wavelength of the outgoing light.

Once the incident light is removed, the luminescence of the fluorescent substance does not disappear immediately, but disappears after a certain period of time.

The fluorescence lifetime has a specific characteristic where the length of the fluorescence lifetime is determined by the temperature, as shown in Figure 2.4. The fluorescence lifetime temperature sensor is a temperature sensor based on this characteristic. At normal temperatures, the afterglow decay time of the fluorescent materials is approximately at the ms level.

Theoretically, the fluorescence decay curve conforms to the single-exponential model:

$$I(t) = AI_p(T)e^{-t/\tau(T)} \tag{2.2}$$

where A is a constant coefficient, t is the afterglow decay time, $I_p(T)$ is the peak fluorescence intensity when excitation is stopped, as a function of temperature T and $\tau(T)$ is

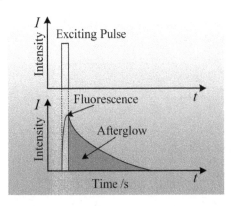

Figure 2.4 Typical fluorescence characteristic curve.

the fluorescence afterglow decay time constant, which is also a function of temperature T, independent of light intensity. The higher the temperature, the more the fluorescence decays and the smaller the $\tau(T)$. Therefore, the temperature decay time constant can be detected in order to know the temperature.

The obvious advantage of the fluorescence lifetime temperature measurement is that the temperature conversion relationship is determined by the single value of the fluorescence lifetime, and is not affected by external conditions such as changes in the excitation light intensity, fiber transmission loss, and coupling loss. The choice of fluorescent materials is a prerequisite for the design of the whole optical fiber fluorescence temperature measurement system [4]. With reference to the selection of fluorescent materials, it is expected to meet several aspects. Firstly, the fluorescent substance can be excited by light of a specific wavelength, where the wavelength of the excitation light and fluorescence radiation cannot be too close in actual applications. Secondly, the consistency of the decay time of the fluorescent material is high and the decay time constant is the same with the same temperature. Thirdly, the fluorescent material should be chemically stable so that long-term exposure to air does not oxidize or corrode, and it has high-grade oil resistance in an application for a power transformer.

2.2.2 Fabrication and Application

A typical fiber optic probe is shown in Figure 2.5. It is mainly composed of optical fiber, fluorescent material, sleeve, heat shrinkable tube, etc. The covering layer is used to protect the optical fiber and the fluorescent material, and, at the same time, the temperature is transmitted to achieve the temperature measurement. The heat shrinkable tube completes the sealing of the optical fiber probe and the fixing of the covering jacket. In general, in order to achieve protection of the probe, the covering layer is generally selected with stainless steel pipes or other non-metallic materials. To ensure the effect

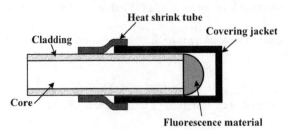

Figure 2.5 Typical design of single fiber fluorescent probe.

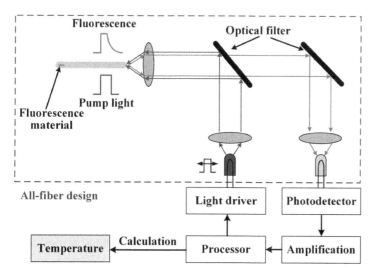

Figure 2.6 Measurement of temperature based on the optical fluorescence lifetime. Source: Reprinted by permission of Jin et al. [6]. © 2020, SPIE.

of heat conduction, its length and thickness are suggested to be controlled and as small in dimension as possible.

Typically, a fluorescent fiber optic temperature sensor is composed of an excitation light source, an optical fiber probe with specific fluorescent material, a photoelectric conversion circuit, an optical filter, an electrical amplifier, a micro control unit (MCU) processor, etc. The overall schematic of the system is shown in Figure 2.6 [5]. A rectangular pulse with an adjustable pulse width is produced by the light source driving the unit to excite the fluorescent material to generate fluorescence. Then, fluorescent signal reflections are converted to a corresponding current signal that is coupled to the photodetector after an optical filter. The fluorescence decay curve data is then amplified and collected by the processor. Finally, the fluorescence lifetime is extracted to provide temperature information. It can be noted that all optical signals are transmitted through the fiber in the design of the fluorescent fiber temperature sensor based on the fluorescence lifetime. The fluorescent temperature sensor does not provide real time measurement as the fluorescence decay curve data needs to be calculated. However, the response is sufficiently instantaneous to meet the requirements of the power transformer application.

It has already been two decades since the first concepts of fiber optic techniques were applied in transformers [6]. It is quite important to facilitate detection by improving the ease of installation into the power transformer tank. An available entire measurement system is composed of the following parts: a temperature monitor instrument (multiple channels), optical fiber probes, a fiber optic bulkhead penetrator, a mounting plate, an external extension fiber, an optional fuel tank wall protection cover, etc. A possible optical fiber temperature measurement system based on multiple fluorescent sensors is depicted in Figure 2.7.

2.2.3 Merits and Drawbacks

As a mature and widely used optical temperature sensor, the fluorescent fiber optic sensor came into use in power transformers very early. With an excellent all-fiber design

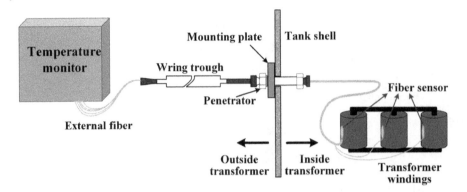

Figure 2.7 Connection illustration of multiple fluorescent temperature measurement system.

and immunity to electromagnetic interference, fluorescent sensors can be immersed on the windings inside the power transformers.

- **Stability and reliability.** The versatile fluorescent material offers stable choices given that the fluorescence decay is an intrinsic property. The fluorescence intensity ratio type is preferable to the fluorescence intensity type, as it is not affected by external factors such as fluctuations in light source and fiber bending.
- **Small size.** Since the typical sensor does not have to be any larger in diameter than the fiber itself, the sensor can, in principle, be extremely small. This allows it to be embedded in or between the windings in a transformer, where size is critical in terms of insulation.
- **Low cost.** As far as optical solutions are concerned, fluorescent material is the most competitive approach given its comprehensive consideration of cost and performance.

However, the shortcomings of fluorescent sensing are also distinctive. It is just a single point measurement and is unable to measure temperature distribution. Although increasing the number of sensors to find the hottest location is probable, it results in complex maintenance and high costs.

On the other hand, the jacket, as a protective layer wrapped around the outermost side of the fiber, has to withstand electrical, mechanical, and thermal stress to ensure normal operation of the fluorescent fiber temperature sensor. Selection of the sheath material not only needs to consider the normal function of the fluorescent fiber temperature sensor but also the chemical stability, heat aging resistance, and mechanical stress resistance of the materials [7, 8].

2.3 FBG-Based Temperature Detection

2.3.1 Detection Principle

Fiber Bragg grating (FBG) is particularly useful for temperature measurement. More than 20 years ago, the FBG sensor was proposed as a promising technique in the electrical power industry due to its capability to be used in high voltage environments [9, 10].

The FBG is an optical device that is formed by exposure to a fringe of UV light. Only the specified wavelength (Bragg wavelength) related to its grating period is reflected for an input light wave from a broadband source [11]. The peak reflected Bragg wavelength λ_B is given by:

$$\lambda_B = 2n_{eff}\Lambda, \tag{2.3}$$

where Λ is the period of grating or index modulation and n_{eff} is the effective refractive index of the fiber. The Bragg grating acts like a mirror that only reflects one very precise wavelength (color). When the optical fiber is strained or when its temperature changes, the reflected wavelength varies proportionally.

For a temperature change ΔT, the Bragg wavelength shift is given by

$$\Delta\lambda_T = \lambda_B(1 + \xi)\Delta T, \tag{2.4}$$

where $\Delta\lambda_T$ is the Bragg wavelength shift and ξ is the fiber thermo-optic coefficient.

The basic principle of an FBG-based temperature sensor is illustrated in Figure 2.8. Moreover, different sensors manufactured using gratings with a specific wavelength can be implemented in series on the same optical line. It is called wavelength-division multiplexing (WDM) technology and is frequently utilized in an FBG sensor, employing several Bragg structure units with different Bragg wavelengths connected in series on to a single optical fiber, as shown in Figure 2.8. Originating from the telecommunication applications, wavelength division multiplexing of FBG for temperature measurement enables a cost-effective solution for a high sensor count that can be used for multi-point or even quasi-distributed measurement. Strictly speaking, the difference between each Bragg wavelength has to carefully considered in order to avoid possible cross interferences and sufficient reflected amplitude to be detected in real applications. Furthermore, to allow for a large number of multiplexed measurements, FBG sensor networks of large capacity can be interrogated using wavelength-division and time-division multiplexing (TDM) or combinations thereof.

By leveraging the WDM technology and strategy, the optical signals are mapped to physical locations along the fiber optic cable. This method enables multi-point sensing along the length of the fiber optic cable inside the power transformer whether a solid-insulating or liquid-insulating type is used.

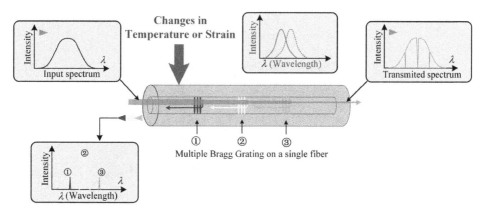

Figure 2.8 Schematic representation of FBG-based temperature sensing and its WDM principle.

At the same time, it should be noted that both changes in temperature and strain will lead to Bragg wavelength shifts. In this sense, the measurement of every parameter should be carried out or compensated for individually in order to get rid of the cross impacts. On the other hand, the sensitivity to a specific parameter can also be enhanced through the effects with the help of some special materials or structures.

2.3.2 Fabrication and Application

Since the bare optical fiber basically works in a temperature range of less than 120 °C, it is not recommended to utilize it in conditions above 100 °C for a long time due to the gratings degradation induced by a high temperature environment. To meet the temperature measurement needs on the hot spot of the transformer windings, a specific heat-resisting material with an appropriate structure is necessary. In addition, strain resistant gratings are also required owing to the FBG sensitivity to strain and temperature. Occasionally, a certain pre-tension is also applied to the FBG while fixing it between the strip and the outer frame to make the sensor more reliable throughout its range. Since the high electric field distribution is in the transformer, the FBG is not supposed to be in contact with any metal body. When the transducer is subjected to the temperature, it results in a tensile axial strain in the FBG [12].

Usually, a bare FBG is specially encapsulated with high temperature resistant materials like polyimide, epoxy resin, polytetrafluoroethylene (PTFE), or Teflon, to ensure a real time heat exchange with windings, anti-corrosion, avoidance of strain interference and insulation defects like partial discharge, or breakdown in high electric field applications.

An available scheme and physical packaging structure of an online FBG temperature probe is shown as Figure 2.9 [13]. An optical fiber jacket is helically incised to ensure heat exchange with insulation oil. Meanwhile, thermal sensors are usually installed between the winding and the cushion block in order to effectively fix sensors without using glue. It has been investigated and verified that it can work for a long time under high temperature in power transformers [14].

On the basis of a single temperature point, the multipoint FBG temperature sensor network (expandable up to tens of probes), integrated inside a power transformer for continuous monitoring of hot-spots on windings, cellulose insulations, and oil, has been further demonstrated and tested with a resolution of 0.1 °C and overall temperature

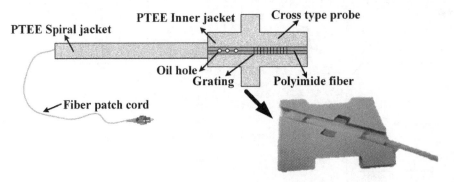

Figure 2.9 Schematic diagram of a typical FBG temperature sensor probe. Source: Reprinted by permission of Yi et al. [13]. © 2016, IEEE.

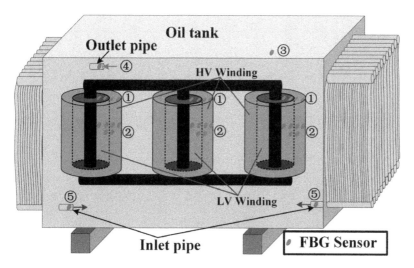

Figure 2.10 Possible locations of the optical FBG sensors.

accuracy of $\pm 1\,°C$. The real time temperature system has a function of history analysis, and helps to achieve online indication of the transformer status and aging behavior [15].

In fact, several guidelines have been found to be beneficial in showing how to arrange the temperature probes inside the transformer. Generally, the upper side of the high voltage winding and the low voltage winding is a good location to monitor the hot spot temperature with consideration of the oil flow (Location 1). Inside the low voltage winding and outside the high voltage winding is another possible location (Location 2) used to compare the temperature in the direction of oil flow and also in the direction perpendicular to oil flow. In addition, installation at the top of the oil tank (Location 3), the outlet pipe (Location 4) where the liquid flows to the radiator, and the inlet pipe (Location 5) helps to profile the temperature distribution [16] (Figure 2.10).

To establish a complete FBG measurement system, the interface, data transmission, communication protocol, etc., should be included as well as the installation of probes. As an example, FBG sensors are installed in a 110 kV oil immersed transformer during its return to a factory repairing period. The measurement position lies in the high voltage windings, top oil, bottom oil, and so on. In this application, the fiber is exported through an interface board and the signals can be connected to an FBG interrogator. The ModBUS and IEC 61850 protocol is built in the interrogator and the data can be transmitted to the server, allowing users to connect remotely and access data systems wirelessly. Moreover, the monitored data can be stored in database files and can be queried and exported conveniently. An intelligent algorithm can be developed into the system to help determine the situation in power transformers. The entire system is depicted in Figure 2.11 [17].

2.3.3 Merits and Drawbacks

FBG is sensitive to temperature change and temperature sensing is one of FBGS's core advantages. FBG is also a cheap and mature optical component for temperature

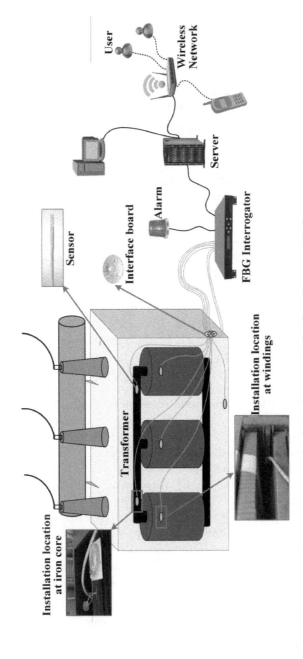

Figure 2.11 Temperature online monitoring system scheme. Source: Modified from Zhang et al. [17].

applications. There are technical benefits associated with the use of fiber optics for hot spot monitoring of transformers.

- **Stability and high maturity.** FBGs have a strong anti-interfere performance and excellent long-term stability. Having been applied in high voltage oil-filled transformers for years, it is considered as one of the most preferred methods of transformer monitoring. An enhanced design and fabrication is beneficial to stabilize engagements in high temperature conditions even higher than 200 °C in power transformers.
- **Real time and high accuracy.** The existing commercial FBG system offers industry level accuracy, precision, and reliability for continuous and real time temperature monitoring of hot spots in oil-filled transformers. In particular, the direct installation within or close to the winding provides an accurate temperature reading.
- **Flexibility to multi-point measurement.** Owing to the WDM function, tens of individual FBG points can be obtained along a single optical fiber.

Alternatively, temperature transmitters can be embedded into a remote terminal unit (RTU), a programmable logic controller (PLC), or other control or monitoring systems to reduce inspection and maintenance costs.

Similarly, the shortcoming of the FBG temperature sensors applied in a power transformer needs penetration into the tank wall and pre-factory installation.

2.4 Distribution Measurement

2.4.1 Quasi-Distributed Temperature Sensing

Since the temperature in a power transformer is always uneven, the HST is the main concern, as mentioned above. Strictly speaking, there are two main approaches to locate transformer hot spots. IEEE Std C57.91 and IEC 354 standards provide an indirect and an empirical calculation based on the traditional top layer oil temperature and the hysteresis effect with consideration of the heat transfer process. However, measurement error is inevitable and the response time is not instant. As a result, the direct method is more preferable with the optical sensors installed in or close to the windings. Nevertheless, it is possible to reside the optical sensor in a very accurate point prior to the installation.

To address this issue, multiple points measurement or distributed temperature detection is necessary to obtain the hot spot zones and points. Especially with the benefits of WDM in the FBG structure, tens of Bragg units share a single optical fiber to measure the temperature at one time. This allows quasi-distributed temperature sensing in a power transformer, as depicted in Figure 2.12. Compared with the point-type conventional winding temperature detection methods, the simple bundle-like quasi-distributed FBG temperature sensors allow a significant reduction in cost per sensor probe. For a mature FBG measurement system, the marginal cost of WDM is negligible, which is important to enable widespread deployment of such real time monitoring techniques in electrical asset applications such as power transformers.

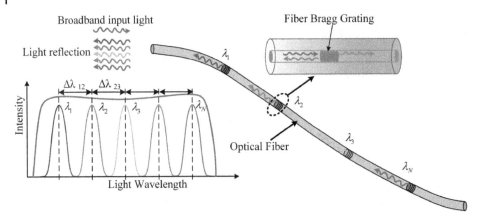

Figure 2.12 Illustration of a multi-point measurement based on FBGs.

2.4.2 Distribute Temperature Sensing

The WDM of FBGs offers a compromised distributed temperature measurement in power transformers. Normally, distributed optical fiber-sensing technology depends on the scattered signal characteristics of the optical signals transmitted in optical fibers to sense temperature or strain along the optical fiber. Therefore, any point along the optical fiber, even as long as kilometer level, can be continuously detected. The distributed temperature sensing (DTS) offers a more accurate, flexible, and easy approach to the temperature profile barely using a fiber in a power transformer.

2.4.2.1 Light Scattering

When light waves propagate inside the fiber, there will be a scattering spectrum of different frequencies. According to the scattered light frequency, it can be divided into elastic scattering (Rayleigh scattering) and inelastic scattering (Brillouin scattering and Raman scattering), as depicted in Figure 2.13 [2]. Rayleigh scattering shares the same wavelength as incident light, since it is induced by the elastically scattered photons. Cased by large scale low frequency vibrational motion of a lattice of atoms, Brillouin scattered photons have a different wavelength. Among them, Raman scattering (including low frequency and high frequency parts) is caused by the energy transfer between fiber molecules and vibrating incident photons. Raman scattering in the high frequency part is found to have strong temperature sensitivity. According to the scattering effect and demodulation topologies, a variety of distributed fiber optic sensing technologies have been utilized in power apparatus, for example, Brillouin optical time-domain reflectometry (B-OTDR), stimulated Brillouin optical time-domain analysis (B-OTDA), optical time-domain reflectometry (OTDR), phase sensitive optical time-domain reflectometry (Φ-OTDR), and Raman optical time-domain reflectometry (R-OTDR) [2, 18–21]. To date, R-OTDR has mainly been adopted in the distributed temperature measurement of power transformers.

2.4.2.2 Raman Based Distributed Temperature Sensing

The simple principle of Raman scattering can be seen in Figure 2.14. Photons with an incident energy of $h\nu_0$ are absorbed by the gas electrons to reach a virtual state from the

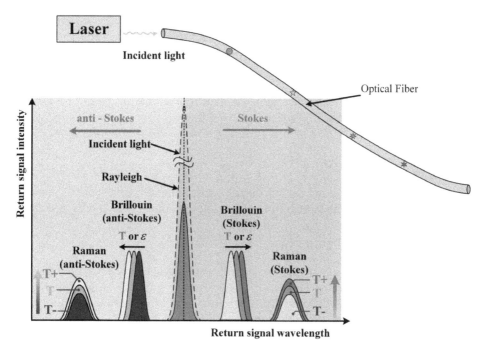

Figure 2.13 Typical spontaneous scattering spectrum in an optical fiber. Reproduced from Chai et al. [2].

Figure 2.14 Schematic diagram of the Raman scattering phenomenon.

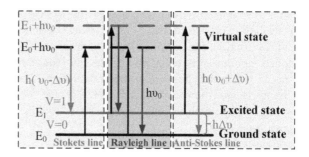

initial ground state. Nevertheless, the virtual state does not possess sufficient energy to maintain the gas electrons. Therefore, a number of electrons return to the ground state and release photons with an energy of $h\nu_0$. This achieves the Rayleigh scattering light without a frequency change. If the electrons return to the vibrational excited state, they release photons with an energy of $h(\nu_0 - \Delta\nu)$ and achieve Raman scattering light (Stokes line) with a frequency change.

When the pulsed light is transmitted through the fiber, a single pulse can simultaneously generate back Stokes Raman and anti-Stokes Raman scattering [22–24]. Suppose P_s is the signal amplitude of the back Stokes Raman scattering and P_{as} the back anti-Stokes Raman scattering; then the ratio of the two signals at the reference temperature T_0 is obtained as

$$\frac{P_{as}(T_0)}{P_s(T_0)} = K_a/K_s(\nu_a/\nu_s)^4 \, e^{-h\Delta\nu/kT_0} \cdot e - (\alpha_0 - \alpha_s)L, \tag{2.5}$$

where K_a and K_s are the cross-sectional coefficients related to the Stokes scattering and anti-Stokes scattering of the fiber, respectively, v_a and v_s are the Stokes and anti-Stokes scattered light frequency respectively, α_0 and α_s are the average loss rate of incident light and Stokes Raman scattered light propagating in the fiber, respectively, L is the distance from the incident end of the fiber to the measured point, h is the Planck constant ($h = 6.626 \times 10^{-34} J \cdot s$), k is the Boltzmann constant ($k = 1.38 \times 10^{-23} J \cdot K^{-1}$), T_0 is the thermodynamic temperature. The specific value of the two-light power at temperature T is calculated as

$$\frac{P_{as}(T)}{P_s(T)} = K_a/K_s(v_a/v_s)^4 e^{-h\Delta v/kT} \cdot e^{-(\alpha_0 - \alpha_s)L}. \tag{2.6}$$

The light source instability and the loss during transmission along the optical fiber can be eliminated. Therefore, the temperature distribution on the optical fiber is given by

$$\frac{1}{T} = \frac{1}{T_0} - \frac{k}{h\Delta V}\left[\ln \frac{P_{as}(T)/P_s(T)}{P_{as}(T_0)/P_s(T_0)}\right]. \tag{2.7}$$

The Raman anti-Stokes signal changes its amplitude significantly with the changing temperature while the Raman Stokes signal is relatively stable. Eventually, the relationship between Raman Stokes and anti-Raman Stokes scattering versus temperature is established and it is an available temperature signal demodulation method for distributed optical fibers.

A distributed temperature measurement method of the transformer winding, based on R-OTDR, is designed and illustrated in Figure 2.15 [2]. A pulse of light is launched into the fiber and a small amount is naturally reflected back along the fiber. The detector used to measure the backscattered Stokes and anti-Stokes band responds over a roundtrip propagation. Both multi-mode and single-mode fibers can be used as the sensing fiber. Single-mode fibers need a high performance laser source and an anti-Stokes photodetector. However, multi-mode fibers may suffer low spatial resolution due to the intermodal dispersion. The position of the temperature change is determined by measuring the arrival time of the returning light pulse. Some special design, such as code correlation technology, enables a single receiver design for both Stokes and anti-Stokes signals, which improves the measurement accuracy and enhances the performance.

The installation of a distributed fiber inside the transformer is also of a great significance. Distributed sensing optical fibers can achieve temperature sensing by attaching to the entire winding surface at the end of each phase winding. The distributed sensing optical fiber is tightly wrapped along the outer surface of the winding; a layer of an

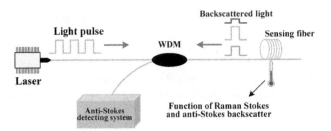

Figure 2.15 The topology and working principle of R-OTDR. Source: Modified from Chai et al. [2].

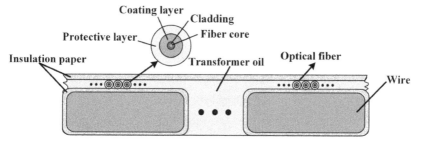

Figure 2.16 Typical distributed sensing fiber based on the winding structure.

insulating paper is tightly wrapped around the wire where the optical fiber has been fixed (Figure 2.16) [8]. During the winding process, the optical fiber itself is stretched while both the pressure and friction of the insulating paper will make it adhere to the wire surface without relative sliding.

To achieve a distributed temperature measurement, a typical R-OTDR-based DTS system is chosen (Figure 2.17). The pulsed light passes through the bidirectional coupler and the fiber-optic wavelength division multiplexer. Two kinds of backscattered light will be sent to the computer for temperature demodulation through the photoelectric conversion and subsequent high-speed analog-to-digital conversion from the avalanche photodiode (APD). Furthermore, an optical tank wall feedthrough is required to allow the fiber into the transformer.

B-OTDR is also a very common technique to achieve DTS in industrial applications. However, its temperature demodulation is affected by both the center frequency and the spectral peak, where errors in the spectral peak fitting reduce the accuracy of the temperature measurement. Moreover, the cost of the B-OTDR technique is relatively high in comparison with the R-OTDR, which is the more preferable solution given that the Raman signal is, theoretically, only impacted by temperature. Therefore, it is essential to comprehensively weigh the pros and cons of every technique in terms of stability, measurement accuracy, and costs to achieve the ultimate solution. Nonetheless, it is noteworthy that B-OTDR and R-OTDR can be combined in a single practical solution [8, 25–27], as shown in Figure 2.18. In such cases, a multi-mode fiber uses R-OTDR to sense temperature information while the single-mode fiber uses B-OTDR to monitor the

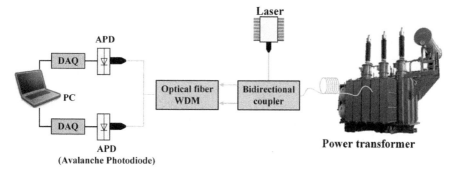

Figure 2.17 Transformer temperature monitoring system based on distributed optical fiber sensing. Source: Adapted from Yunpeng et al. [8].

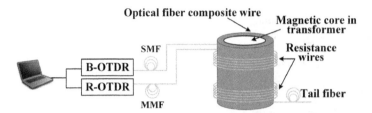

Figure 2.18 The optical fiber winding composite model and the measurement system.

winding process in real time (possible mallet knocking, sharp corners, etc.) to determine whether the sensing fiber is twisted or broken.

Apart from R-OTDR, Raman-based optical frequency domain reflectometry (R-OFDR) is also available for DTS using a continuous wave (CW) laser with different frequencies instead of a pulsed laser. The duty cycle of the pulsed laser is a very important parameter to determine the spatial resolution in OTDR; however, there are no duty cycle requirements for the OFDR technique. The full temperature and spatial resolution are achieved even for rather long single-mode fibers [20].

Temperature profiles can be obtained through two possible OFDR-based configurations. The simple approach is by measuring the amplitudes and phases of the anti-Stokes and Stokes methods, where the resultant ratios can be used to determine the temperature profiles. However, as with the more complex approach, an electrical spectrum analyzer (ESA) is used to record the amplitude and phase of the scattering signals, which are then post-processed by the inverse Fast Fourier transform (IFFT) algorithm.

It is preferable to perform OTDR for a long distance (up to tens of km) with the pulsed laser and have OFDR for a short distance with the CW laser, employed by amplitude modulation [21]. However, for power transformers, both R-OTDR and R-OFDR are applicable when the spatial resolution, sensor range, and temperature resolution are taken into consideration.

2.4.2.3 Rayleigh-Based Distributed Temperature Sensing

As mentioned above, Brillouin or Raman scattering responds to temperature, and is available for DTS with the help of a time/frequency domain reflectometer. However, the spatial resolution is limited to the meter level and is difficult to improve due to the extremely low scattering intensity, whereas Rayleigh scattering in the backscattered model provides an alternative approach.

Although Rayleigh scattering is not excited nor influenced by temperature, Rayleigh backscattering is caused by defects that induce a local variation in the permittivity. Specifically, the optical fiber is slightly and spatially stretched and compressed by the change of temperature or strain, which makes a difference in proportional frequency shifts. The analysis can be performed by small sections, whose signals are transformed into the frequency domain with regard to the Rayleigh pattern. To achieve a significantly higher spatial resolution, a tunable scanning laser is necessary, and a coherent optical frequency domain reflectometer is usually adopted to only measure the perturbations if the polarization matches that of the reference. Resolution of the Rayleigh OFDR can be even as good as a millimeter range [28, 29].

Similarly, the installation and configuration of the OFDR-based sensing system has been utilized and instrumented on the transformer core in real applications [30]. The sensing fiber is mounted on the four faces of the racetrack-shaped transformer core (front, inside, back left leg, back right leg, and outside) as well as inner/outer loops through coating the fiber with thermally conductive paste and sticking the paste-coated fiber directly to the core surface. The distributed temperature measurement is then verified and evaluated. This method is used to minimize pre-strains in the fiber, while maximizing the thermal conductivity between the core and fiber.

2.4.3 Merits and Drawbacks

FBG is able to achieve quasi-distributed sensing with limited test points but varying spatial resolution with regard to the arrangement of the Bragg gratings. Alternatively, by using the fiber as a sensing medium, distribution techniques are utilized along the entire length of the fiber rather than at discrete points that provide an excellent approach to distributed temperature monitoring of power transformers. With the help of distributed temperature data, the precise and visible mapping of the temperature in a power transformer is available and is not limited to hot-spot temperatures. The Raman scattering-based sensing technology can realize a DTS along the entire optical fiber, where R-OTDR is selected as the monitoring topology to meet the requirements of real-time hot-spot locations and transformer internal temperature field measurements.

The merits of distribution measurement are obvious, but there are several shortcomings that need to be considered.

- **Low spatial resolution.** Distributed sensing is possible with the above techniques, but the spatial resolution is still as low as the meter level, which could lead to possible measurement errors.
- **Pre-factory installation.** Since the distribution measurement relies on the inside installation of the optical fiber, the purpose of the distribution sensing mainly focuses on the pre-factory tests or researches instead of volume production and application. Although fibers are intrinsically for safety of insulation, the introduction of the long fibers and fixing structure is still deemed to be a threat to the long-term operation of power transformers.
- **Excessive costs.** At present, the devices of these techniques are generally very expensive and it iss not possible for them to be mounted as in routine online monitoring. Even the cost of the optical measurement system may be higher than the distribution power transformer itself.

2.5 Conclusion

Undesirable heat causes a temperature rise typically in the windings in power transformers. The increasing temperature is responsible for the degradation of oil/paper insulation and increases the probability of thermal fault occurrences. Once an efficient method is applied to accurately and safely monitor the temperature, it is more important to control an overload situation, which reduces the risk of impacting the transformer's lifetime.

The optical solution used as a direct method provides accurate measurements of the hot spots in power transformers. In summary, mainly three typical forms of optical sensing are introduced and discussed in this chapter (Figure 2.19), including point

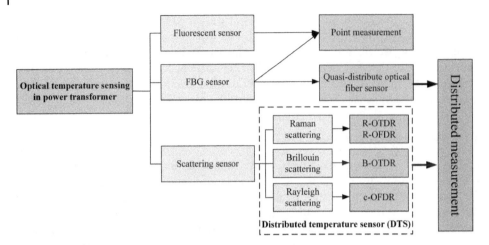

Figure 2.19 Optical temperature sensing techniques in a power transformer.

measurement for HST in transformer windings and DTS for mapping the temperature distribution in a transformer.

Novel optical fiber techniques have been developed to prevent a "fever" in power transformers, with the merits of small sizes, an easy-to-use configuration, immunity to electromagnetic fields, rapid response for real-time monitoring, and the ability to be embedded in transformers. Therefore, an optical temperature provides an insightful and effective perspective to power transformers. However, most of the optical temperature sensing units are pre-installed in a laboratory scenario and need to undergo more evaluation or examination with long-term online operations. Currently, the cost of the various interrogators and optical components still cannot be neglected. It is believed that a fiber optics temperature monitoring system, with a mature industrious design and installation, provides added value to various types of power transformers, that is, it does not only provide long-term thermal information but it also reduces inspection, lowers maintenance costs, and improves real-time transformer loading capability.

References

1 Betta, G., Pietrosanto, A., and Scaglione, A. (2001). An enhanced fiber-optic temperature sensor system for power transformer monitoring. *IEEE Transactions on Instrumentation and Measurement* 50 (5): 1138–1143.

2 Chai, Q., Luo, Y., Ren, J. et al. (2019). Review on fiber-optic sensing in health monitoring of power grids. *Optical Engineering* 58 (7): 072007.

3 Arabul, A.Y., Keskin Arabul, F., and Senol, I. (2018). Experimental thermal investigation of an ONAN distribution transformer by fiber optic sensors. *Electric Power Systems Research* 155: 320–330.

4 Zhao, Y., Chen, M., Lv, R. et al. (2016). Small and practical optical Fiber fluorescence temperature sensor. *IEEE Transactions on Instrumentation and Measurement* 65 (10): 2406–2411.

5 Jin, X., Yang, J., and Pan, L. (2020). Design of fluorescent fiber temperature sensor based on fluorescence lifetime. In: *Sixth Symposium on Novel Photoelectronic Detection Technology and Application*. SPIE.

6 McNutt, W.J., McIver, J.C., Leibinger, G.E. et al. (1984). Direct measurement of transformer winding hot sport temperature. *IEEE Transactions on Power Apparatus and Systems* PAS-103 (6): 1155–1162.

7 Shuguo, G., Jun, G., Jing, P. et al. (2017). Thermal aging of the fluorescence optical Fiber temperature sensor sheath materials in transformer oil. *High Voltage Engineering* 43 (8): 9.

8 Yunpeng, L., Xinye, L., Huan, L. et al. (2020). Development of 35 kV oil-immersed transformer with built-in distributed optical Fiber. *High Voltage Engineering* 46 (6): 9.

9 Rao, Y.J. (1999). Recent progress in applications of in-fibre Bragg grating sensors. *Optics and Lasers in Engineering* 31 (4): 297–324.

10 Picanço, A.F., Martinez, M.L.B., and Rosa, P.C. (2010). Bragg system for temperature monitoring in distribution transformers. *Electric Power Systems Research* 80 (1): 77–83.

11 Hirayama, N. and Sano, Y. (2000). Fiber Bragg grating temperature sensor for practical use. *ISA Transactions* 39 (2): 169–173.

12 Mamidi, V.R., Kamineni, S., Ravinuthala, L.N.S.P. et al. (2016). High-temperature measurement using fiber Bragg grating integrated with a transducer. *Optical Engineering* 55 (11): 116104.

13 Yi, J., Shuang, L., Li, X., and Wei, L. (2016). Fiber Bragg grating sensors for temperature monitoring in oil-immersed transformers. In: *2016 15th International Conference on Optical Communications and Networks (ICOCN)*, 1–3. IEEE.

14 Wei-gen, C., Jun, L., You-yuan, W. et al. (2008). The measuring method for internal temperature of power transformer based on FBG sensors. In: *2008 International Conference on High Voltage Engineering and Application*, 672–676. IEEE.

15 Ribeiro, A.B.L., Eira, N.F., Sousa, J.M. et al. (2008). Multipoint fiber-optic hot-spot sensing network integrated into high power transformer for continuous monitoring. *IEEE Sensors Journal* 8 (7): 1264–1267.

16 Kweon, D.-J., Koo, K.-S., Woo, J.-W., and Kwak, J.-S. (2012). A study on the hot spot temperature in 154kv power transformers. *Journal of Electrical Engineering and Technology* 7 (3): 312–319.

17 Zhang, X., Yao, S., Huang, R. et al. (2014). Oil-immersed transformer online hot spot temperature monitoring and accurate life lose calculation based on fiber Bragg grating sensor technology. In: *2014 China International Conference on Electricity Distribution (CICED)*, 1256–1260. IEEE.

18 Ma, G., Wang, Y., Qin, W. et al. (2020). Optical sensors for power transformer monitoring: a review. *High Voltage* (to be published).

19 Razzaq, A., Zainuddin, H., Hanaffi, F., and Chyad, R.M. (2019). Transformer oil diagnostic by using an optical fibre system: a review. *IET Science, Measurement and Technology* 13 (5): 615–621.

20 Hill, W., Kübler, J., and Fromme, M. (2010). Single-mode distributed temperature sensing using OFDR (EWOFS'10). In: *Fourth European Workshop on Optical Fibre Sensors*. SPIE.

21 Silva, L.C.B.d., Pontes, M.J., and Segatto, M.E.V. (2017). Analysis of parameters for a distributed temperature sensing based on Raman scattering. *Journal of Microwaves, Optoelectronics and Electromagnetic Applications* 16: 259–272.

22 Hausner, M.B., Suárez, F., Glander, K.E. et al. (2011). Calibrating single-ended fiber-optic Raman spectra distributed temperature sensing data. *Sensors* 11 (11): 10859–10879.

23 Meng, L., Jiang, M.-s., Sui, Q.-m., and Feng, D.-j. (2008). Optical-fiber distributed temperature sensor: design and realization. *Optoelectronics Letters* 4 (6): 415–418.

24 Soto, M.A., Signorini, A., Nannipieri, T. et al. (2011). High-performance Raman-based distributed fiber-optic sensing under a loop scheme using anti-stokes light only. *IEEE Photonics Technology Letters* 23 (9): 534–536.

25 Liu, Y. et al. (2018). A feasibility study of transformer winding temperature and strain detection based on distributed optical fibre sensors. *Sensors* 18 (11): 3932.

26 Liu, Y., Yin, J., Fan, X., and Wang, B. (2019). Distributed temperature detection of transformer windings with externally applied distributed optical fiber. *Applied Optics* 58 (29): 7962–7969.

27 Liu, Y., Yin, J., Tian, Y., and Fan, X. (2019). Design and performance test of transformer winding optical fibre composite wire based on Raman scattering. *Sensors* 19 (9): 2171.

28 Loranger, S., Gagné, M., Lambin-Iezzi, V., and Kashyap, R. (2015). Rayleigh scatter based order of magnitude increase in distributed temperature and strain sensing by simple UV exposure of optical fibre. *Scientific Reports* 5: 11177.

29 Samiec, D. (2012). Distributed fibre-optic temperature and strain measurement with extremely high spatial resolution. *Photonik International* 1: 10–13.

30 Lu, P. et al. (2019). Real-time monitoring of temperature rises of energized transformer cores with distributed optical Fiber sensors. *IEEE Transactions on Power Delivery* 34 (4): 1588–1598.

3

Moisture Detection with Optical Methods

Moisture in the air makes the breath quite comfortable and fresh, but it is a great threat for power transformer insulation, which can even be seen as a slow but deadly poison for oil-immersed power transformers. Typically, the insulation system in a transformer mainly consists of presspaper and transformer oil. The increase in moisture content in the insulation system accelerates the deterioration of the oil/paper insulation and decreases the life span of the insulation by half. In addition, moisture content is also prone to dramatically decrease the inception voltage of partial discharge (PDIV), it can increase the risk of bubble evolution at high temperatures, and it can even create electrical breakdown and sundry problems. Moreover, bubble formation increases in the insulation system while the water content increases, leading to potential discharges. Therefore, the vicious circle continues. A high moisture content in insulation oil is definitely undesirable; therefore, all moisture content in a transformer is thoroughly dried up before the start of the operation. Normally a transformer is strictly sealed and hardly contains any moisture, but water cannot be completely removed considering the atmospheric moisture ingress and decomposition processes under conditions of electrical or thermal faults, or fairly long period of operation.

In this sense, the measurement of moisture in transformer insulation is gaining importance and is becoming beneficial for large-scale power transformers, especially for the transformers that operate for considerable periods of time. In particular, those are reaching or surpassing their fixed lifespan, which are more prone to have a wet insulation threat in service. Therefore, the detection of moisture in transformer oil is still gaining importance with regard to the state-based maintenance of oil-immersed transformers.

3.1 Online Monitoring of Moisture in a Transformer

Moisture management in power transformers is a concern. Whatever the source of the water might be, during the dynamic operation of a transformer, a kind of moisture distribution or migration equilibrium is established to some degree within the oil. Therefore, moisture determination of the insulating liquid/presspaper is adopted as a routine measurement for transformers and related electrical equipment to investigate the cellulosic insulation status. Nowadays, the detection of moisture in transformer oil can be mainly categorized as offline detection and online detection types. The offline monitoring method traditionally includes distillation, chromatography, gravimetric method,

Optical Sensing in Power Transformers, First Edition. Jun Jiang and Guoming Ma.

and chemical technique based on Karl-Fischer titration (KFT) [1]. It is usually the custom to absorb water from the surroundings and get the water content to the laboratory to be tested. The detection period is therefore always too long to obtain an instant decision and so is the subject of speculation. In anticipation of real-time measurement to obtain a proper profile of a specific transformer, online monitoring of moisture in the transformer deserves more attention and development.

3.1.1 Distribution of Moisture in the Power Transformer

Specifically, the concentration of water less than 1000 mg /kg is emphasized in the scenarios of moisture content in a power transformer, which is quite different from the usual unit of relative moisture saturation (RH, %) in the air versus the condition or type of insulation oil. Up to now, the conventional mg/kg (parts per million, ppm) measurements have been well established and with reference data readily available for condition monitoring. There is a motivation to convert the new measurement data of RS into mg/kg equivalents and vice versa. Also, mg/L is used to describe the moisture content in some literature and the conversion can be done with the proportional oil density constant.

The insulation system is established with insulation oil and paper/pressboard [2–5].

As to the paper and pressboard elements, which provide the mechanical stability of the insulation, their main component is cellulose. Water contained in the cellulose insulation is a vital consideration as it exponentially increases the aging rate. The basic unit of the cellulose is the glucose molecule and these molecules can be linked to form a cellulose chain, as shown in Figure 3.1. Cellulose is a polymer compound with a degree of polymerization (DP) at around 1200 for new paper, indicating that there are on average 1200 glucose molecules in a cellulose chain. There is no doubt that the large chains give the paper the mechanical strength it requires to fix the windings, even in rough conditions such as when a short circuit has occurred. Also, the large number of unsaturated hydroxyl groups is related to the moisture content in a power transformer, since water is a key component in the mechanism (hydrolysis reaction) of cellulose degradation.

Most electrical transformers utilize refined oils distilled from crude oil sources as insulating fluids. In nature, transformer oil is a mixture of many hydrocarbon molecules with different molecular weights, operating under high pressure, high temperature, and strong electromagnetic fields in long-term running, especially stressed by the catalytic action of various additives in oils. The oil is very easy to be oxidized, bringing in some polar products and precipitated water. The polymer compounds in the oil are also oxidized and cracked to generate water. At the same time, the oxidation of the oil accelerates

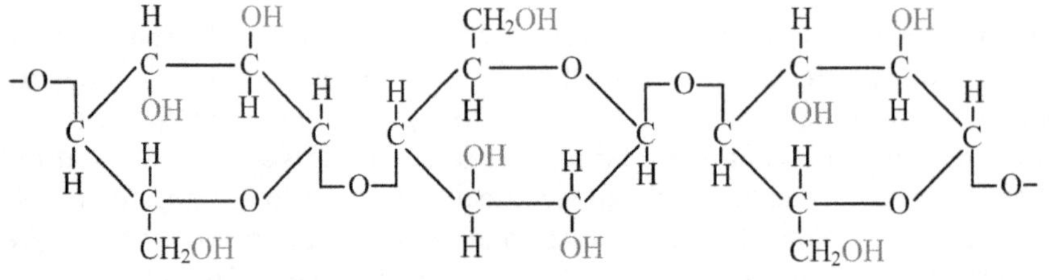

Figure 3.1 Molecule structure of cellulose consisting of glucose rings with unsaturated OH groups.

the aging of the oil and the insulating paper. Generally, there are three main forms of moisture in transformer oil: dissolved, emulsified, and free water. Because water is composed of polar molecules, since the oil has strong hydrophobicity, dissolved water is quite confining to be dissolved in the oil. The presence of dissolved water in the transformer oil will increase the acid value of the oil, reduce the oxidation stability, accelerate the aging and deterioration, and reduce the breakdown voltage performance of the oil, causing partial discharge and other problems. The presence of the emulsified state also reduces the breakdown voltage and increases the dielectric loss. Free water in the oil will cause certain damage to the structural stability of the oil molecules, bringing in a reduction of the insulation performance of the oil.

Although some water remans in a suspension state and cannot be absorbed by the paper in the insulation system in a power transformer, actually, most of the water is in the form of dissolved water and is available to move from oil to the solid insulation (presspaper) as the transformer progresses toward equilibrium. At equilibrium, the water dissolved in the oil is at an equilibrium with the water adsorbed in cellulose and vice versa; as a result, relative saturations of both components of the insulation system (oil and paper) are equal. Therefore, special equilibrium curves are developed to establish the relationship between the absolute water content in oil and the water content in paper to evaluate the moisture content in the insulation. Since temperature is very important to the thermal equilibrium status, it is suggested that a relative saturation measurement should be provided as well as the measurement of the oil temperature at the location of the moisture sensor. Benefitting from the two parameters, the relative saturation can be converted in absolute water content in oil in ppm.

Oil temperature is the main influencing factor of the existing form of water. With the change of oil temperature, the three existing forms of water can convert to each other. However, as the temperature of the oil increases, the solubility of the oil in water will increase, as shown in Figure 3.2. Typically, the solubility of water in the transformer oil is temperature dependent, varying from 20 mg/kg at 0 °C to 800 mg/kg at 100 °C [6]. No matter what kind of water exists in the oil, it causes the physical and chemical properties

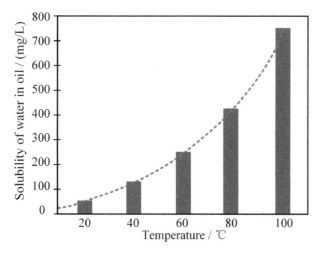

Figure 3.2 Water solubility in transformer oil is dependent on the temperature.

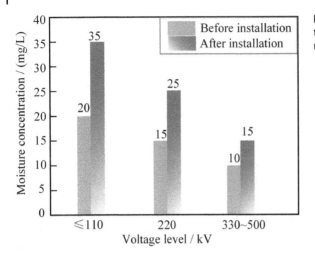

Figure 3.3 Quality standard of transformer oil with regard to moisture content.

of the oil to change, deteriorates the insulation, and easily causes operation failure of the transformer.

According to the related standard *Quality of transformer oils in service* (GB/T 7595-2008), the permissible moisture level in paper is inferred from values of the water content in oil, which should be less than 35 mg/L with regard to 330–500 kV oil-immersed power transformers, as illustrated in Figure 3.3.

In the recent IEEE Std C57.106 – 2002, the permissible moisture level in paper is inferred from values of the water content in oil, assuming thermal stability and moisture equilibrium between the paper and oil. In practice, when the moisture content of the transformer oil approaches 50 mg/kg at 20 °C, the oil is either reconditioned or replaced since the dielectric breakdown of the transformer oil is significantly increased when the water content rises above 50% saturation. The generally accepted specification for the water concentration of this reconditioned or new oil is ≤10 mg/kg [7].

3.1.2 Typical Moisture Detection Techniques

The moisture content of solid insulation is a concern for power transformers as it causes several harmful effects on the insulation system. As a traditional technique to measure the water content of solid transformer insulation, the Karl Fischer test is widely accepted in the laboratory, although it is recognized that a single sample cannot reveal the moisture content in paper if the oil temperature is unstable. As to the application in a power transformer, direct access to transformer winding is in anticipation of collecting the paper samples, which is rarely possible. Moreover, the water content of the sample on its delivery to the test laboratory is prone to vary as it may adsorb water from the atmosphere or along with the errors arising from unfavorable and non-standard sampling. In other words, an online moisture content assessment is expected.

Online monitoring systems are available and involve continuous recording, which allows the integration of temperature variations and the computation of a dependable value for the moisture content in paper. To address the moisture issue literally, the CIGRE working group A2.30 published a report on the measurement of moisture within transformer insulation systems in 2008 [8]. Since then, online monitoring of the water content has attracted more attention, and various online moisture detection probes have sprung up, especially since an instant and continuous monitoring data is necessary to get and evaluate a tendency moisture profile. In this profile, a dielectric frequency response analysis is a powerful tool used to determine the water content in power transformers and to assess their condition and remaining life span based on the polarization depolarization current (PDC) measurement or frequency domain dielectric spectroscopy (FDS). With the development of further automation, the measurement time can be drastically reduced and, as a result, the whole measurement and assessment process makes the techniques easy and reliable for users. However, it is still not a good candidate for the field application in the long term. In 2018, as a continuation of the work of CIGRE Brochure 349, "Moisture equilibrium and moisture migration within transformer insulation systems," and new insights into the distribution of moisture between insulating liquids and solid insulation is introduced. The CIGRE working group D1.52 further emphasizes the moisture measurement and assessment in transformer insulation and recommends the moisture capacitive sensors for online application [9]. Availability of continuous moisture-in-transformer measurement by means of capacitive sensors and its relation to temperature opens up new diagnostic possibilities in comparison to conventional KF spot measurements.

In the last decade, capacitive polymer sensors have been increasingly used to evaluate the moisture content in power transformers. Basically, a capacitive moisture sensor is composed of parallel plate capacitors. At least one of the electrodes is permeable to water vapor and allows water molecules to diffuse into the dielectric polymer layer. In theory, the absorbed water molecules increase the permittivity, which can be measured as increased capacitance of the sensor element. It is necessary to guarantee that the sensor is very selective to water, but is insensitive or has little response to other molecules or products in oil. Typically, a parallel plate capacitive humidity sensor is fabricated on properly cleaned glass substrates, having a thin metal film of a few micrometers thickness, which acts as a lower conducting electrode. The basic structure of the parallel plate capacitive sensor and its possible installation are shown in Figure 3.4 [10].

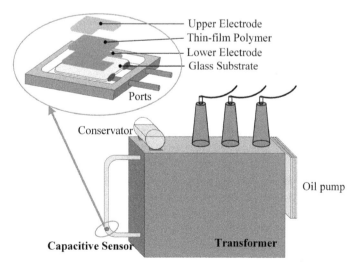

Figure 3.4 Schematic structure of a parallel plate capacitive sensor. Source: Reprinted by permission from Islam et al. [10]. © 2020, IEEE.

In particular, the capacitive thin film polymer sensor provides an effective approach to the moisture measurement in transformer oil based on the vapor pressure of water dissolved in insulating oil. Its application enables access to a thorough observation of moisture dynamics in a high voltage insulation. It is helpful to estimate the paper moisture level based on the mathematical models like that of Fessler and Piper to calculate the water activity level and make it come true with real time and continuous monitoring and observation [11, 12].

It is natural to understand that the only moisture at the surface of the sensor can be detected, and it is important to avoid installing it in stagnant oil since the moisture fluctuations may not be collected. Installing the probe at the correct location is important and should be in the path of oil flow. Not only is the installation position strictly limited, but its process is quite dependent on the diffusion and temperature. Capacitive sensing technique is easily affected by the environment, which limits the use in occasions where high-precision measurement is in anticipation. A temperature compensation is required for an estimation of the water content of paper to be valid, as the water activity probe cannot be installed close to a high voltage winding due to the risk of compromising the insulation and causing dielectric breakdown [13]. Besides, with regard to a high voltage in-situ application, the capacitive sensors have to face the risk of electromagnetic interferences.

Most of the currently available sensors measure the oil outside of the transformer with regard to the harsh conditions (compact and high electric field). Due to the limitations of current techniques, it is important to explore new technologies and probes that can be utilized to measure the moisture content of paper insulation to cope with future challenges. Emerging fiber optic sensors are ideally suited for these harsh conditions, where the optical fiber is both compact and insensitive to a high electric field. Fiber optic-based sensors have been used over many years to measure winding temperature, and thus a similar technology to measure the insulation moisture content would be highly beneficial. Their miniature size, light weight, flexibility, immunity to electromagnetic interference, and ruggedness make them an ideal sensor device for moisture detection in power transformers. Mainly three optic fiber moisture sensors are involved to this application in the chapter: fiber Bragg gratings (FBGs), evanescent wave types (mainly micro/nano fiber (MNF)), and an interferometric Fabry–Perot (FP) structure.

3.2 FBG-Based Moisture Detection

3.2.1 Detection Principle

As mentioned in Chapter 2, an FBG sensor operates analogously to an optical filter, where most wavelengths of light pass through freely while certain wavelengths are reflected, called the Bragg wavelength. Therefore, the concern for moisture measurement by FBG is how to make the most of the sensitive parameters through special design and fabrication. The difficult point of the sensor design is how to relate the shift of Bragg wavelength to the moisture content.

Ideally, the moisture content of the oil directly next to the insulation paper would be measured for an accurate estimation of the paper water content. However, the objective is to develop an FBG-based sensor that is specifically target moisture. A hygroscopic

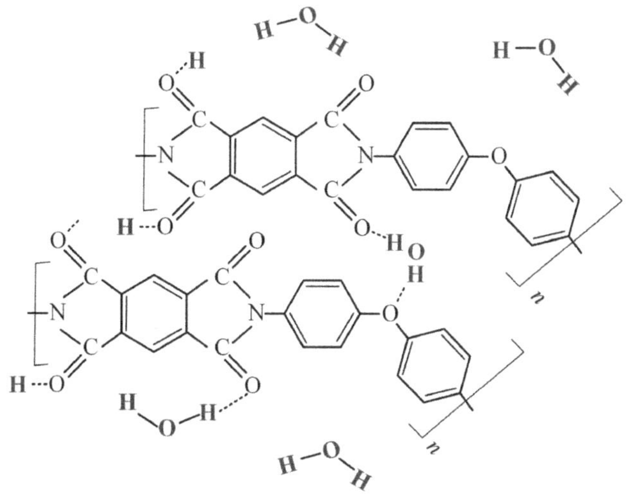

Figure 3.5 Repeat units of a PI molecule and the possible sites for water. Source: Reprinted by permission of Melcher et al. [17].

material is necessary to establish the relationship between the moisture content and Bragg shifts; that is, the Bragg grating can be coated with a suitable polymer material to absorb surrounding moisture and is then characterized with respect to the moisture and temperature level. Polyimide (PI) is the most common material due to its low cost and stable performance. Polymethyl methacrylate (PMMA), polyvinyl acrylate (PVA), di-Ureasil (di-U600), carbon nanotube (CNT), and graphene are also excellent candidates for hygroscopic materials to be used as the coating reacting with water due to their effective hygroscopic properties [14–16].

To illustrate the principle of an FBG-based moisture sensor, polyimide coating is considered as the hygroscopic material. Figure 3.5 illustrates the molecular structure of polyimide [17]. Two different sites are likely to have a bound water molecule: the first one is the oxygen of the ether linkage and the other one is the four carbonyl groups. The possible sites for water molecules are indicated schematically. At low humidity levels, the sorption of water at the carbonyl groups is much more likely than the sorption at the ether oxygen. Thus, the key is the kinetic moisture transport due to the solubility interactions and/or polymer "swelling" of the membrane. In the end, a diffusion balancing state is achieved at the surface boundary region of the polyimide.

As an instance, Figure 3.6 presents the typical structure of a moisture sensor with a certain thickness of PI coating around it. First of all, a polymer coated FBG gives an initial dry Bragg reflection, then the fiber is exposed to humidity, and lastly the polymer coating swells after absorbing water, straining the underlying Bragg grating and shifting the Bragg reflection to a longer wavelength. Indeed, it is necessary to mount the coating just around the gratings zone and put pressure on the Bragg gratings only to have a sensing effect.

To get an insight into the principle, a standard Bragg grating strain and temperature relationship model is established first. The relative change in wavelength is listed in terms of the strain ε_z on the fiber along the optic axis. The change in temperature ΔT

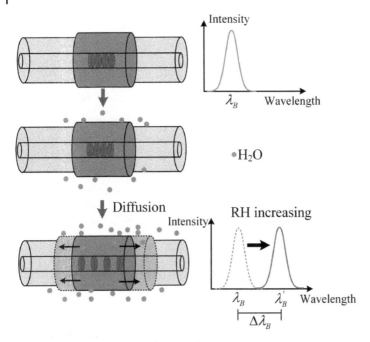

Figure 3.6 FBG-based sensors for humidity monitoring.

can be expressed as

$$\frac{\Delta \lambda_B}{\lambda_B} = (1 - p_e)\varepsilon_z + ((1 - p_e)\alpha_\Lambda + \alpha_n)\Delta T \tag{3.1}$$

in which p_e is the effective strain optic coefficient, α_Λ is the thermal expansion coefficient of the fiber, and α_n is the thermo-optic coefficient of the fiber.

Since a humidity sensitive polymer coating around Bragg gratings only generates strain along the fiber axis, the increase of moisture content m in transformer oil can be detected by wavelength shifts. In this scenario, the adapted Bragg grating strain and temperature relationship in a moisture sensor can be expressed as

$$\frac{\Delta \lambda_B}{\lambda_B} = (1 - p_e)\varepsilon_{rh} + ((1 - p_e)\alpha_T + \alpha_n)\Delta T \tag{3.2}$$

where ε_{rh} and ε_T mean the water-absorbing coating induced strain on the fiber due to the moisture content variation Δm and temperature changes ΔT, respectively. In terms of these changes, ε_{rh} and ε_T can be expressed as

$$\varepsilon_{rh} = \left(\frac{A_p E_p}{A_p E_p + A_f E_f} \right)(k_{m1} - k_{m2})\Delta m \tag{3.3}$$

$$\varepsilon_T = \left(\frac{A_p E_p}{A_p E_p + A_f E_f} \right)(k_{T1} - k_{T2})\Delta T \tag{3.4}$$

where A_p and A_f are the cross-sectional areas of the polymer and fiber respectively, E_p and E_f are the Youngs' moduli, k_{m1} and k_{m2} are the coefficients of moisture expansion, and k_{T1} and k_{T2} are the coefficients of thermal expansion of the polymer and fiber, respectively. Here k_{m2} is assumed to be zero due to the fact that glass material in an optical fiber is resistant to water absorption under any circumstances.

With regard to constant temperature in a specific surrounding, Eq. (3.2) can be rewritten as Eq. (3.5), which is then substituted in Eq. (3.3) to give the expression of Eq. (3.6):

$$\frac{\Delta \lambda_B}{\lambda_B} = (1 - p_e)\varepsilon_{rh} \tag{3.5}$$

$$\Delta \lambda_B = \lambda_B (1 - p_e) \left(\frac{A_p E_p}{A_p E_p + A_f E_f} \right) k_{m1} \Delta m \tag{3.6}$$

where k_{m1} is a measure of how much a material swells with humidity (moisture), and, therefore, the larger the coefficient the larger the degree of the material swelling. With the experimental data of the coefficient of moisture expansion of the polymer, the slope of a linear plot of $\Delta \lambda_B$ versus Δm can be confirmed [18].

3.2.2 Fabrication and Application

Recent publications have been researching various polymer coatings as moisture-sensitive layers for FBGs. The main goal is to develop a sensor through optimizing the coating thickness to achieve the required specificity, accuracy, and sensitivity.

The key to an FBG moisture sensor is the control of the coating of water-absorbing material. It is easy to know that humidity sensitivity can be increased by increasing the coating thickness since a thicker layer of hygroscopic material enables a higher strain effect. In addition, the interface between the water and the polyimide increased with the polyimide layers and the water absorbance also increased.

However, experiences show that any variation in coating thickness places a varying strain over a certain Bragg grating, causing the Bragg reflection peak to become distorted. If this distortion is significant, then tracking the peak during measurements can be difficult. The FBG coated with one thick polyimide layer was easily bent in the high temperature curing process of fabrication, inducing an uneven layer and a multi-peak FBG spectrum.

To address the difficulty of practical FBG moisture sensors, several key points should be considered.

- A multiple-layer structure is a good choice to fabricate the coating instead of a single thick layer. Five or more layers are recommended to be prepared in order.
- Certain pre-stress helps to develop the coating. FBG can be vertically placed and a light force (not to destroy the fiber) is applied on the free end of FBG in the curing process.
- Silane coupler, or a similar material, is also beneficial to strengthen the connection of each layer.

The above-mentioned measures have been proved to improve the property of an optical fiber humidity sensor, even on a specially designed FBG sensor proposed for detecting equivalent salt deposit density (ESDD) on insulators of the power transmission line [19].

For instance, a 10-layer polyimide film around the Bragg grating section is formed by dip-coating repeated 10 times. The sensors are re-coated with the hygroscopic material (polyimide solution, PI-2560) to get a coating with an estimated thickness of 35 µm.

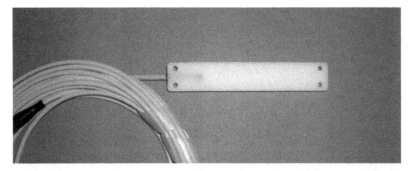

Figure 3.7 Photograph of a fabricated moisture FBG sensor with a PI coating and the packaged version.

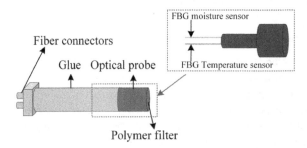

Figure 3.8 Schematic diagram of the packaged sensor probe with moisture and temperature measurements.

Then, the sensor is packaged in a specific plastic box formed with a 3-D printer, which is air permeable and allows the moisture to penetrate the cover and arrive at the sensor [20]. In this manner, a packaged, 10-layer polyimide-coated FBG humidity sensor was developed, as shown in Figure 3.7.

Eventually, a temperature compensation is necessary to be done since FBG is sensitive to temperature variations. In order to eliminate the temperature-induced wavelength shift from the moisture sensor, a temperature-only FBG sensor is embedded into the probe. A further example of utilizing a PI-coated FBG moisture sensor is illustrated in Figure 3.8. Prior to an on-site application, the performance should be assessed in the laboratory under experimental conditions of controlled wetting and drying cycles [21, 22]. This approach has the merit of a compact design and a minimal invasive volume.

The fixing and installation of a moisture sensor is also an issue. Since the components of a large-scale power transformer are complex, the location or purpose of the measurement is important to select the position. Whatever the target is, it is important to confirm that the oil measured is truly representative of the oil inside the transformer tank. The intention is to interpret the paper moisture or to see whether there is a high relative moisture saturation in the oil system.

There are several possible locations that can be considered for the installation of moisture detection in a power transformer:

- The cooling circulation pipe after radiators/coolers. It is an optimal place to assess the risk of decreased dielectric strength due to high relative moisture saturation of the oil.
- The position of valves. It is a very easy way to merge the sensors into a valve with a non-destructive manner. However, a cutout zone may form during the sensing area, which is unfortunate for obtaining an accurate measurement.

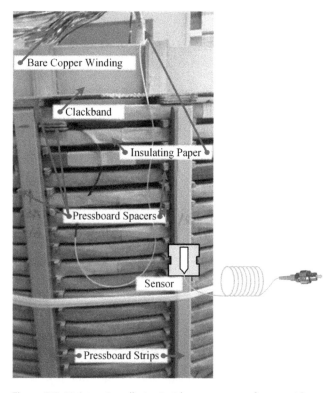

Bare Copper Winding

Clackband

Insulating Paper

Pressboard Spacers

Sensor

Pressboard Strips

Figure 3.9 Moisture installation inside a power transformer with a customized pressboard space.

- Circulating oil path. It is crucial that the sensor is positioned in the oil flow, especially in the case of bypass installations. The use of moisture sensors in the flow of hot and cold oil will provide additional useful information.

As an example, Figure 3.9 shows the installation of a packaged FBG moisture sensor in a high voltage transformer, in which customized pressboard spacers are good for sensor package embedding and providing extra mechanical protection to the sensor sensitive head [23].

Moreover, power transformers undergo daily, weekly, and seasonal cyclic loading patterns regularly and fiber optic sensors are capable of reflecting the moisture dynamics in the power apparatus. The variation in the moisture measurement is likely to have been caused by their different locations and corresponding temperatures. Therefore, the flexibility offers a new insight for a maintenance worker to manage the utilities better during normal and overloading conditions.

It is easy to understand that the moisture distribution in a transformer is not only a function of temperature but also depends upon the location and moisture gradient in a large power transformer insulation tank. To get an accurate and complete moisture profile in the tank, or the insulation system, a distributed fiber optics moisture sensing array is recommended since FBG has the function of wavelength division multiplexing (WDM) to monitor the water content at multiple points. In addition, the temperature can be detected simultaneously in a series chain. An example is depicted in Figure 3.10 [13]. The distribution measurement system includes a light source, sensing FBG chains, a data acquisition unit, and a data processing unit. Various digital filtering techniques can

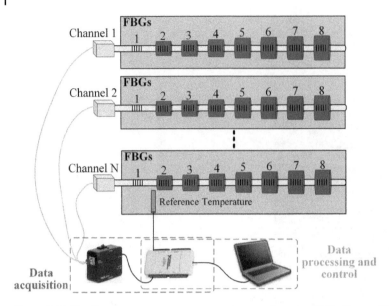

Figure 3.10 Distribution moisture measurement system based on FBGs. Source: Modified from Ansari et al. [13].

be explored to promote the sensors' stable output. Every FBG chain can be connected to a channel with multiple moisture and temperature sensing probes. With the strength of immunity to electromagnetic interference, the sensors can be virtually introduced anywhere inside a transformer. Along with the fiber chains, the moisture readings from this array will be compared with other measurements to ascertain effectiveness and any moisture gradients within the system. As a result, the application of these distributed moisture measurements helps to provide more precise moisture measurement and location information inside a transformer.

3.2.3 Merits and Drawbacks

Although moisture is not included in the sensitive parameters of conventional FBGs, it still provides a link and conversion between the strain and the moisture content through specific design and materials. Especially, a kind of polymer-coated optical fiber sensor is analyzed in theory, and the fabrication of coatings, sensor installation inside a real transformer, and the distribution form are presented to prove the effectiveness in the section as well. Many advantages have been revealed through the complete solution using the FBG technique.

- **Highly sensitive.** Humidity sensors are common, but the moisture content in insulation oil is a vital parameter, and the detection level is much lower than the usual needs for daily life or industrial applications. Moreover, the accuracy of the FBG-based sensor is acceptable to power transformers and shows close performance of non-optical moisture probes to detect a small amount of water.
- **Internal installation.** Different from non-optical conventional probes, immunity to electromagnetic interference and small size opens up the possibility to install the sensors virtually anywhere inside a transformer. It also provides closer interaction with

the possible wet part in a transformer, which is beneficial to get an accurate moisture measurement.

- **Distribution measurement.** Since a large scale power transformer occupies a large space and needs a huge amount of insulation oil, the flexibility of inside installation and WDM offer new insights for distribution measurement of moisture content to be able to manage the moisture mapping better during normal and abnormal conditions.

Additionally, there are some challenges and concerns to be researched and investigated in the near future. Investigations on using polymer-coated FBG sensors have been presented to measure the water content in transformer oil. The common thought on this work is to utilize the specific coatings on the optical structure of FBG due to the water-induced swelling of the fiber and change of refractive index (RI). Unfortunately, the chemical coatings may not be stable with the long-term immersion in oil and have the risk of being dissolved. The durability of some hydrophilic polymeric materials has been questioned as they can be corroded by the hot water and oil after long-term use. Therefore, the long-term stability of these sensors needs to be tested.

On the other hand, FBGs are sensitive to both strain and temperature and therefore the cross-impact may bring on some fluctuations or inferences. In addition, the wavelength demodulation device for an FBG is slightly expensive and complex, limiting industrial applications and large-scale manufacturing.

3.3 Evanescent Wave-Based Moisture Detection

3.3.1 Detection Principle

Experiences show that the square of the refractive index (RI) is represented as the relative permittivity of an optical material. Since water and insulating oil have different values of refractive index, this means that different moisture contents in oil lead to a variety of refractive indices as a whole. Accordingly, the moisture content in a transformer oil sample can be transferred into the change in refractive index of the sample. However, if the light is restricted inside the fiber, then it would be impossible to sense and react to the change of ambient RI and moisture content as well. Thus, it is necessary to "leak" the light from the fiber through the form of an evanescent wave.

In general, there are mainly three approaches to form the evanescent wave effect with optical fibers.

- **Bending of fiber** [24, 25]. It is well established that a fiber should not be bent since the situation introduces a degree of power loss. However, that inspires a good idea to carry out measurements. Once the fiber is in a sharp bend or macro-bend with a radius of curvature exceeding the critical radius, the higher order modes radiate out of the fiber first, resulting in loss or attenuation since higher modes are bound less tightly to the fiber core than the lower order modes. Thus, light rays are lost in the cladding influenced by the ambient RI, which results in power loss and thus attenuation in accordance. As an example, a U-bent fiber configuration, shown in Figure 3.11, can be used to detect the moisture content. The bare portion of the fiber is then made into a bend of fixed radius of curvature, which constitutes the optical sensor probe (OSP). Furthermore, a sensing film can be attached at the surface of the bare fiber core to

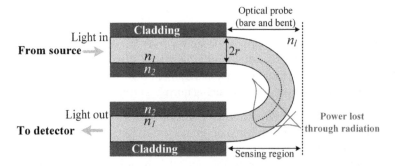

Figure 3.11 Geometry representation of the radiation loss of an evanescent wave at a fiber bend.

enhance the interaction of the evanescent field with the sensing film and surrounding moisture.

- **D-shaped fiber** [15, 16]. Apart from bending the fiber, it is also possible to enlarge the effective detecting area of the sensing fiber and make the energy of the evanescent wave field interact with the analyzed substance (like RI based moisture) directly. The sensitivity of the sensor can be improved greatly by increasing the grinding depth of the D-shaped optical fiber. D-shaped fibers enable the application of chemical and biochemical evanescent field sensing [26].
- **Micro/Nano fiber (MNF)** [27–31]. MNF, a kind of emerging optical fiber, has enormous prospects in optical fiber sensing due to the large evanescent field. With this consideration, MNF is mainly analyzed both from the theoretical and operational aspects in this chapter.

To give a simple explanation, a micro-nano fiber can be stretched from a normal single-mode fiber (SMF) with a thinner diameter. Usually, there are transition and uniform regions, respectively, with a tapering fabrication, as illustrated in Figure 3.12. Moreover, MNF can be seen as the fiber core and surroundings outside the fiber of the cladding. Then the light interacts with the substance to be detected and the relationship between adhered particulates and the loss of MNF can be theoretically modeled and calculated.

According to electromagnetic theory, when light is totally reflected at the interface between the core and the cladding, although no energy leaks out of the core, there is also an electromagnetic field in the cladding that is an evanescent field. It decreases exponentially with depth across the interface. It is assumed that a lossless passive optical

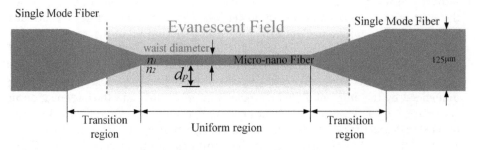

Figure 3.12 Evanescent field generation and distribution around the MNF.

waveguide suits the MNF, so the Maxwell equation can be simplified to the Helmholtz form to express its physical properties.

Basically, the fundamental mode occupies the small diameter MNF (at the micrometer level or less) and thus the transfer equation solution is deduced to be

$$\left\{ \frac{J'_1(U)}{UJ_1(U)} + \frac{K'_1(W)}{WK_1(W)} \right\} \left\{ \frac{J'_1(U)}{UJ_1(U)} + \frac{n_2^2 K'_1(W)}{n_1^2 WK_1(W)} \right\} = \left(\frac{\beta}{kn_1} \right)^2 \left(\frac{V}{UW} \right)^4$$

(3.7)

where J_1 is the first Bessel function and K_1 is the modified second Bessel function. Then

$$U = \frac{d\sqrt{k_0^2 n_1^2 - \beta^2}}{2}$$

(3.8)

$$W = \frac{d\sqrt{\beta^2 - k_0^2 n_2^2}}{2}$$

(3.9)

$$V = \frac{k_0 d\sqrt{n_1^2 - n_2^2}}{2}$$

(3.10)

where $k = 2\pi/\lambda$ is the wave number, λ acts as the transmission wavelength, d is the diameter of MNF, n_1 and n_2 are the core refractive indices and the dielectric cladding refractive index, and the undetermined coefficient β tightly relates the optical field transmission around MNF. In Figure 3.12, d_p is the depth of the MNF evanescent field. It can be seen from the above equations that when changing the refractive index of surrounding substances (insulation oil), the output optical power change is in accordance with the absorbance in the evanescent field.

Typically, the refractive index of the standard pure transformer oil (Oakley 25# transformer mineral oil) $n_2 = 1.4616$ at room temperature and $n_{water} = 1.333$. Then the oil with moisture has a slightly lower RI value, and the moisture content has an influence on the RI reading. As to MNF with diameters of 0.8, 3, 5, and 10 µm, the optical field energy distribution is shown in Figure 3.13.

For the MNF with diameters of 1–50 µm, the evanescent field energy ratio can be seen in Table 3.1.

The diameter of MNF dominates the distribution of an evanescent wave, that is, a small one is beneficial to achieve the large proportion of the evanescent wave distribution

Table 3.1 Evanescent field energy ratio of MNF with different diameters.

No.	Diameter of MNF (µm)	Evanescent field energy ratio (%)
1	1	94.2
2	5	21.0
3	10	0.80
4	30	0.075
5	50	0.0089

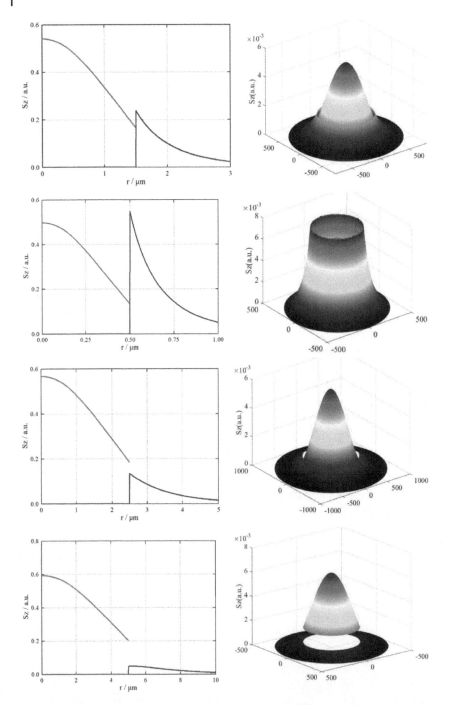

Figure 3.13 Evanescent wave energy distribution diagram of MNF with different diameters. (a) Evanescent wave energy distribution diagram of 800 nm MNF. (b) Evanescent wave energy distribution diagram of 3 μm MNF. (c) Evanescent wave energy distribution diagram of 5 μm MNF. (d) Evanescent wave energy distribution diagram of 10 μm MNF.

and more sensitive interaction with the substance. To establish the relationship between concentration and the evanescent field, the light power loss should be calculated.

According to the Beer–Lambert Law,

$$P_{out} = P_{in}e^{-\alpha L} \tag{3.11}$$

where P_{in} is incident light power, P_{out} is output optical power, L represents optical path length, and α indicates the absorption coefficient, relating to the Fresnel transmission coefficient of light (T) and the number of times of light reflection per unit length of the fiber (N). Therefore

$$\alpha = NT \tag{3.12}$$

At the interface of the MNF and the transformer oil, the Fresnel transmission coefficient equals

$$T = \frac{k0n_2 \cos \theta}{(n_1{}^2 - n_2{}^2)\sqrt{\sin^2\theta - \sin^2\theta_c}} \tag{3.13}$$

where θ is the fiber light incident angle and k_0 is the constant coefficient. All the parameters are constant except n_2, which is variable; thus the change of the output optical power depends only on the refractive index of the oil sample. Therefore, P_{out} can be expressed as a function of the refractive index n_2 of the transformer oil sample:

$$P_{out} = f(n_2) = e^{\left[\ln(P_{in}) - \frac{n_2 kcLN \cos\theta}{(n_1{}^2 - n_2{}^2)\sqrt{\sin^2\theta - \sin^2\theta_c}}\right]} \tag{3.14}$$

Ideally, the RI of the oil sample is related to the moisture content without any other factors. In this way, the numerical relationship between the moisture content in the transformer oil and the additional loss of MNF is theoretically established.

3.3.2 Fabrication of MNF

At present, there has been limited research on the fabrication of micro-nano optical fibers. The common methods that have been reported are the chemical etching technique and the melting taper drawing method [30]. The melting taper drawing method can be divided into the one-step taper drawing method, the two-step taper drawing method, and the sapphire taper block material method. The main difference between these methods is that different heating sources (like the high temperature flame, laser, and solid heater) are used to heat different raw materials (such as ordinary SMF, polymer solution, and block glass material) to prepare the micro-nano fiber. These methods have their own advantages and disadvantages. The main factors affecting the sensing effect of the micro-nano fiber are the diameter, diameter uniformity, surface smoothness, etc. The preparation of the micro-nano fiber with a small diameter, good diameter uniformity, and good surface smoothness can improve the sensing effect.

3.3.2.1 Chemical Etching Method
The principle of chemical etching is that hydrofluoric acid reacts with silicon dioxide to corrode the optical fiber, which makes the diameter of optical fiber smaller. The existing research shows that the corrosion rate of hydrofluoric acid solution at room temperature

is closely related to the concentration of hydrofluoric acid solution [32]. Thus, different concentrations of hydrofluoric acid solution and the corrosion time can be controlled to prepare micro-nano fibers with different diameters. Generally, the concentration of hydrofluoric acid solution, the corrosion time of hydrofluoric acid, and the used amount of hydrofluoric acid are comprehensively determined according to the required corrosion length and the depth of the micro-nano fiber.

There are some disadvantages related to the fabrication of the chemical etching method:

1. The diameter uniformity is not very good, which is caused by the slightly different corrosion rate to the fiber center in all directions
2. It is difficult to prepare a long micro-nano fiber.
3. 3, The surface roughness of a micro-nano fiber is usually large, which affects the sensing performance of a micro-nano fiber.

3.3.2.2 Fused Biconical Taper Method

Flame-Brushing Technique The flame-brushing technique is a common method for fabrication of a micro-nano fiber. The typical device of one-step taper method is shown in Figure 3.14.

Micro-nano fibers with diameters ranging from hundreds of nanometers to several microns can be fabricated by the one-step taper method. The method has many advantages, such as a simple experiment device, low cost, good repeatability, an easy-to-build experiment platform, etc. The disadvantages of this method are mainly due to the instability of the flame when the flame is heated, and the difficulty of fine control of the flame temperature and size, which makes it difficult to prepare a micro-nano fiber with a small diameter. Because the flame has a certain air impact force, when the diameter of the micro-nano fiber is small, sometimes it will be broken by the impact force of the flame. Therefore, when the diameter of the micro-nano fiber is small, the ceramic heating plate can be used for heating. The one-step taper method has been improved to achieve the fabrication of the micro-nano fiber with low cost and convenient operation.

Self-Modulated Taper-Drawing Technique The self-modulated taper-drawing technique is one of the most common methods used for the fabrication of micro-nano fibers, which

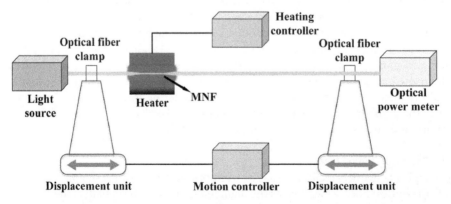

Figure 3.14 The experimental setup for fabrication of a micro-nano fiber by the flame-brushing technique.

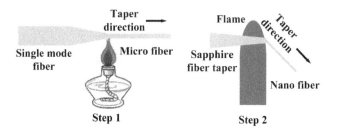

Figure 3.15 The fabrication of a micro-nano fiber by self-modulated taper-drawing technique.

was firstly proposed by Tong et al. [33]. The diagram of the two-step drawing process of a micro-nano fiber is shown in Figure 3.15. In the first step, a standard SMF is pre-tapered to a microfiber using a conventional flame-pulling technique. Then in the second step, a sapphire fiber taper with the pre-tapered microfiber is drawn to form a nanofiber. The diameters can be as low as tens of nanometers.

The advantage of the self-modulated taper-drawing technique is that it can fabricate a micro-nano fiber with small diameters. However, due to the influence of the flame of the alcohol lamp and the surrounding air, it is easy to be broken, and the repeatability of the test is not efficient when the diameter of the micro-nano fiber is small. The operation requirements in the process of taper drawing are high, which makes the actual operation more difficult

Direct Drawing from Bulk Technique The above methods use SMF as the raw material to prepare a micro-nano fiber. In order to enrich the selection of raw materials, a method of fabricating a micro-nano fiber from bulk glass materials is also proposed [15, 34]. The experimental implementation diagram of the fabrication of micro-nano fiber by Sapphire taper block material method is shown in Figure 3.16.

The main disadvantage of this method is that it is difficult to control the diameter of the micro-nano fiber precisely because of the randomness of the deformation when the block glass material is tapered, which leads to the poor repeatability of the experiment.

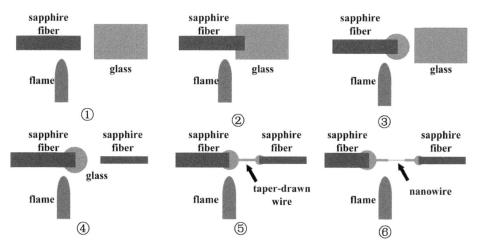

Figure 3.16 The implementation diagram of the fabrication of micro-nano fiber by direct drawing from bulk.

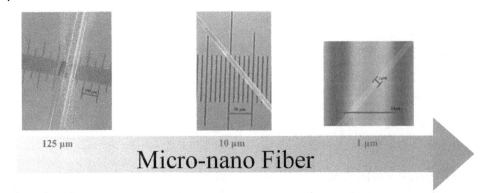

Figure 3.17 Enlarged MNF images with different diameters.

In order to obtain the short taper, different fabrication techniques should be comprehensively compared and selected to get the suitable sensing MNF probes (Figure 3.17).

3.3.3 MNF Moisture Detection

There is no existing MNF moisture sensor in a power transformer. However, some experiments have been carried out in the laboratory [22, 23, 35]. A typical experimental setup is illustrated in Figure 3.18, where a normal multimode fiber and MNF are used at the same time. It is easy to obtain the light power loss in the path of the MNF sensing arm, and the moisture content in the oil tank can be evaluated according to the theoretical model of moisture content.

Apart from the simple configuration, a type of evanescent wave absorption sensor is constituted with the Sapphire fiber and electrode to detect real-time water concentration below 150 mg/l, as shown in Figure 3.19 [35]. There are three main parts for the optical sensor: photodetector (right), a thermal emitter and bandpass filter (left), and an oil flow-through cell (middle). Although it is slightly difficult to implement in practical

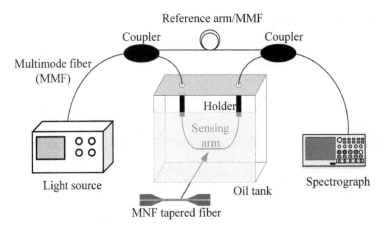

Figure 3.18 Experimental setup to detect moisture content based on an evanescent wave of MNF.

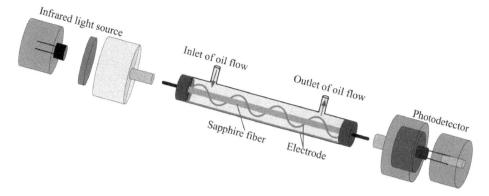

Figure 3.19 Fiber optic evanescent wave absorption sensor. Source: Modified from Holzki et al. [35].

applications with multiple components, it provides a good reference for future industrious design.

However, it should be noted that the relationship between the moisture content and the power loss is ideal, since the refractive index of the liquid is influenced by many factors, such as temperature, impurities, degradation, turbidity, etc. Thus, if a precise measurement of moisture is based on the RI reading, the scenario needs to control the other factors accurately or with necessary compensation. Occasionally, the RI can provide an estimate on degradation of the insulation liquids as well, and it is also considered as good practice to evaluate the health status of the transformer oil in this manner. Apart from the evanescent field of the optic fiber [36], aging phenomenon of transformer oil can be detected by the Mach–Zehnder interferometry technique [37] and FBG scheme [38] as well.

In addition, the oil level is available to be detected through the evanescent field [39, 40], the FBG technique [41], or the multi-mode interference (MMI) effects [42], which is very important to be kept within a specific limit for reliable running of a transformer.

3.3.4 Merits and Drawbacks

Based on the establishment of the evanescent wave, which is mainly based on MNF, its distribution is strongly related to the refractive index, which has a relationship with the moisture content. Following that, an MNF-based optical sensor is used to establish an affinity between the moisture in oil and the evanescent field distribution. MNF provides a superb alternative for humidity-sensing applications in power transformer oil with several distinctive merits.

- **All-fiber sensing and simple structure.** The sensing probe is very simple and can be done through ordinary SMF with an all-fiber microstructure. Without any coating or polymer material, it has no effect on the transformer oil and the optical signal is not influenced by a complex electromagnetic environment in the field. As a consequence, the MNF sensor is free from air composition and has a low cost.
- **High sensitivity.** The diameter is the critical factor to determine the evanescent field distribution and the sensitive response to moisture variation. According to existing research, the MNF with a diameter less than 1 μm has a comparable performance

in sensitivity and dimension compared with both existing fiber sensors and mature products. Additionally, its sensitivity can meet the actual needs of the power transformer.

- **Real-time response.** Since the measurement does not need any separation or additional procedures, the installation is very simple. That is to say, the response of moisture detection is very instant with the data of optical power loss from the input and output. The MNF sensor can be installed and dipped directly in the field transformer tank to achieve online monitoring.
- **Degradation evaluation of oil.** The refractive index in transformer oil is easily influenced by impurities and change in color of the degraded oils. Therefore, it can be used to evaluate the entire aging degree instead of the measurement of moisture content on that occasion.

In spite of these advantages, several factors must be further considered for field applications. In fact, the key to measuring the moisture in oil is to sense the refractive index value, which is more susceptible to the properties or factors of oil itself, like temperature and aging. Considering the factor of temperature, the response of MNF to moisture in oil will be influenced to some limited degree and calibration work needs to be carried out to compensate for the temperature variation prior to being directly applied on the spot. In other words, it is necessary to scale the measurement prior to the moisture in oil test to get a more accurate evaluation when using MNF sensors.

3.4   Fabry–Perot-Based Moisture Detection

3.4.1   Detection Principle

The Fabry–Pérot interferometer (FPI) is another important component in optical sensing, which can also be used for moisture content detection and measurement. An optical cavity with two parallel reflecting surfaces (and with it the resonance length) easily forms the interferometric phenomenon and then optical waves pass through it. More importantly, recent advances in the fabrication technique and novel materials allow the creation of very precise tunable FPIs. On the other hand, traditional bulk optic components such as beam splitters, combiners, and objective lenses have been rapidly replaced by small-sized fiber devices, providing the possibility of micro-scale applications in fiber scales.

Optical fiber interferometers have been very successful in sensor applications of humidity. Typically, three styles of FPI-based moisture sensors are provided in Figure 3.20a to c [22, 43]. Accordingly, the wavelength shifts of the sensor upon exposure to an environment of varying relative humidity (concentrations of Level 1–Level 4 increase in turn) are illustrated. To trace water or humidity concentration, humidity-induced swelling stretches the fiber or cavity and creates an unbalanced path length to get the distinctive output, as shown in Figure 3.20a. In this manner, the FP-based moisture sensors have a similar measurement mechanism as FBG moisture transducers.

Also, a sensing layer can be coated directly on the outermost end (or simply sandwiched in between two partially reflecting mirrors), as depicted in Figure 3.20b. The thickness of the cavity is related to the wavelength of the input diode laser source and

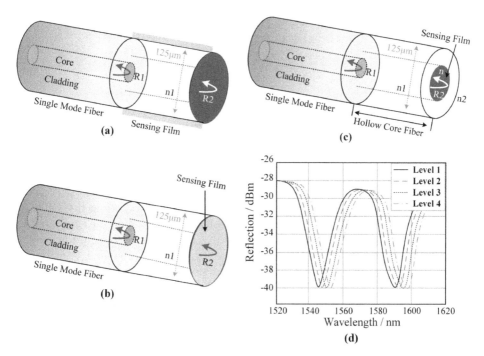

Figure 3.20 (a) Sensing coating wrapped around the fiber, (b) sensing at the end face of the fiber, (c) hollow core fiber as the cavity, and (d) wavelength shifts of the sensor in different moisture contents. Sources: Alwi et al. [22]; Yeo et al. [43].

can be optimized to match the operation in a good condition. It should be noted that the refractive index of the sensing material on or inside the cavity is supposed to have a dependence on humidity. Therefore the resonance is shifted in response to the humidity change and can be conveniently detected by performing intensity measurements at a fixed wavelength. However, it is unavoidable that it reacts with the temperature change, and the structure suffers from cross-sensitivity to temperature, which can be corrected using a suitable compensation scheme.

Moreover, another form is based on the micro-structured fiber and a section of hollow core fiber is spliced to an SMF and coats the tip of the hollow core fiber with sensing film. Then an interesting interferometric moisture sensor is created, as illustrated in Figure 3.20c. Similarly, an optical path modulation with a change in ambient moisture content leads to the swelling effect of the sensing material and the output signal can be detected and monitored, as shown in Figure 3.20d.

3.4.2 Fabrication and Application

In addition, there are many derivative products on the basis of FPI. On one hand, a Fabry–Perot cavity and two identical uniform FBGs can be combined to form a new sensor for simultaneous measurement of humidity and temperature, as shown in Figure 3.21. Exposure to the moisture and its variation, the water-absorbing polymer coating (i.e. polyimide) expands and stretches the fiber, inducing strain on the FBG and on the FP cavity. The induced strain alters the grating period, cavity length, and

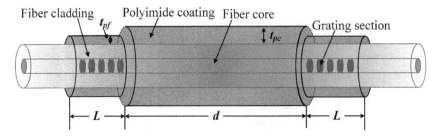

Fiber cladding   Polyimide coating Fiber core

t_{pf}

t_{pc}

Grating section

L   d   L

Figure 3.21 Layout of an FBG-FP sensor for simultaneous measurement of moisture and temperature.

effective refractive index of the fiber. At this situation, it is easy to understand that if the water-absorbing polymer coating (i.e. polyimide) is uniform along the FBG and the cavity, the strain and temperature response on the FBGs and on the cavity are the same. Therefore, wavelength shifts can be obtained but the peaks remain unchanged, making the simultaneous measurement of the moisture content and temperature impossible. As a result, the key is to make coatings on the FBG and the cavity with different thicknesses, with the thickness on the area between FBGs (t_{pc}) larger than that of the FBGs (t_{pf}). In turn, the strain induced on the fiber cavity is higher than that on the grating section. Thus, the cavity serves as a humidity sensor, and the FBG acts as a temperature sensor. Through the delicate design, a change in the intensity of the peaks beside the spectrum shift due to the different strain responses, temperature, and moisture measurement are verified and obtained at the same time [44].

Apart from FBG humidity sensors, long period gratings (LPGs) are the alternative optical fiber gratings (OFGs) and are based on choices to be employed as moisture sensing applications. Similarly, hygroscopic coatings are employed to absorb and desorb vapor in response to humidity changes and achieve a humidity sensing purpose. In contrast, LPG-based sensors are required to connect both sides of the fiber optic. From this perspective, FBG has an obvious advantage due to its one-end measurement. However, the combination of LPG and FPI provides a possible configuration to work as the reflection modality, as shown in Figure 3.22. The forward propagation path and propagation path of the reflection facilitate the use of this configuration for a potential commercial device [14, 45]. A sensing material positioned at the fiber tip or a mirror can be set to detect the reflection band shift and intensity attenuation. This solution allows new scenarios of LPG and FPI applications.

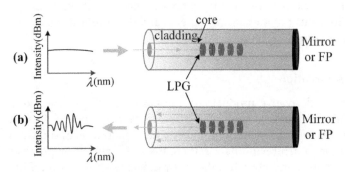

core

cladding

Mirror
or FP

(a)   Intensity(dBm) λ(nm)   LPG

(b)   Intensity(dBm) λ(nm)   Mirror
or FP

Figure 3.22 Combination of LPG and FPI.

3.4.3   Merits and Drawbacks

Similarly, FPI provides another ideal approach to moisture measurement with the advantages of easy alignment, high coupling efficiency, and high stability with regard to the in-line structures. The great strength is the flexibility of FPI, offering a hybrid solution of grating plus interferometric configurations.

The strategy involves a combination of both fiber grating and interferometric configurations to achieve more effective moisture measurements than through the use of either approach alone. Since the conventional structure is sensitive to temperature change, the hybrid solution helps to eliminate the detection error from the actual moisture sensing results. Moreover, the simultaneous measurement of the temperature of moisture is feasible. Additionally, the current trend of miniaturization in fiber optic interferometers enables the sensors to operate inside the transformer tank.

As for the drawbacks of hybrid FPI-based sensors, it is obvious to find the complex configuration of the system and the cavity itself is not very reliable during the fabrication and installation. Reliability and integration should be the consideration for future improvement.

3.5   Conclusion

Moisture is a threat to the insulation of oil insulated power transformers. It increases the risk of breakdown of paper insulation and accelerates the aging of the cellulose insulation as well. There are several ways to get water, even if a transformer is leak-free. Water is released as a by-product of the normal transformer aging process. Moisture content in a transformer is regarded as one of the major factors in diagnosing its conditions. Additionally, moisture management is very important to old transformers widely operating in many countries today. Thus, moisture has a detrimental effect on transformer performance and it is, therefore, of great importance to carry out continuous online monitoring during normal and overload conditions.

Since the paper insulation within a transformer is usually inaccessible, it is feasible to detect the moisture content in oil to represent the moisture level. The introduction of capacitive polymer sensors opens up a real time moisture measurement of transformer insulating oil, although electromagnetic interference and risk of installation are still challenging. Although optical moisture sensors are researched and designed in a laboratory stage (only a few types have been installed), fiber optic-based sensors have the advantage of being both online and immune to high electromagnetic fields. Due to the merits of intrinsic safety, low cost, compact volume, light weight, and low transmission loss, optic-based detection offers a suitable technical solution for online monitoring of moisture. This chapter provides a comprehensive outline and status quo of optical moisture sensing techniques, broadly categorized as FBG, evanescent wave (mainly MNF), and Fabry–Perot interferometric-based sensors.

Specifically, with the water-induced swelling effect by polymer coating, moisture-in-oil measurement based on FBG is achievable. Its flexibility of inside installation and WDM feature offer distribution measurement of moisture content to get a better management of the moisture mapping during normal and abnormal conditions. As a representative of the evanescent wave technique, emerging MNF makes full use of the evanescent wave distribution to sense the change of moisture content through RI, even without any additional coatings. It is both a simple and all-fiber structure. It is also possible to evaluate the degradation level of insulation oil in a power transformer. With regard to the FPI type, its flexibility provides possibilities of hybrid solution together with FBG or LPG to simultaneous moisture and temperature measurement.

There are negative consequences to high moisture levels that are inside the transformer tank. The effect of moisture on thermal aging and reduction of dielectric strength are concerns for aging transformers. Optical techniques are available to continuous moisture-in-oil measurement and obtain a proper moisture profile of a transformer with flexible placement. However, monitoring moisture in oil based on a single physical or chemical performance index is incomplete and redundant, and it is impossible to obtain comprehensive and accurate information. Therefore, a multi-variable feature fusion technique is necessary to construct a scientific, reasonable, and practical online moisture monitoring for power transformer status.

References

1 Bin, C. and Ge, L. (2020). Research progress in on-line monitoring methods of micro-water content in transformer oil. *High Voltage Engineering* 46 (4): 12.

2 Martin, D. and Saha, T. (2017). A review of the techniques used by utilities to measure the water content of transformer insulation paper. *IEEE Electrical Insulation Magazine* 33 (3): 8–16.

3 Sarfi, V., Mohajeryami, S., and Majzoobi, A. (2017). Estimation of water content in a power transformer using moisture dynamic measurement of its oil. *High Voltage* 2 (1): 11–16.

4 Mahanta, D.K. and Laskar, S. (2018). Water quantity-based quality measurement of transformer oil using polymer optical fiber as sensor. *IEEE Sensors Journal* 18 (4): 1506–1512.

5 Bian, C., Wang, J., Bai, X. et al. (2020). Optical fiber based on humidity sensor with improved sensitivity for monitoring applications. *Optics & Laser Technology* 130: 106342.

6 Margolis, S.A. and Angelo, J.B. (2002). Interlaboratory assessment of measurement precision and bias in the coulometric Karl Fischer determination of water. *Analytical and Bioanalytical Chemistry* 374 (3): 505–512.

7 "IEEE Guide for Acceptance and Maintenance of Insulating Mineral Oil in Electrical Equipment," *IEEE Std C57.106-2015 (Revision of IEEE Std C57.106-2006)*, pp. 1–38, IEEE, 2016.

8 C. A2.30, "Moisture equilibrium and moisture migration within transformer insulation systems," CIGRE, 2008.

9 C. D1.52, "Moisture measurement and assessment in transformer insulation – Evaluation of chemical methods and moisture capacitive sensors, CIGRE, 2018.

10 Islam, T., Tehseen, Z., and Kumar, L. (2020). Highly sensitive thin-film capacitive sensor for online moisture measurement in transformer oil. *IET Science, Measurement & Technology* 14 (4): 416–422.

11 W. A. Fessler, W. J. McNutt, and T. O. Rouse, "Bubble formation in transformers: Final report,"; United States, Technical Report, EPRI-EL-5384, 1987. https://www.osti.gov/biblio/6301241.

12 Piper, J.D. (1946). Moisture equilibrium between gas space and fibrous materials in enclosed electric equipment. *Transactions of the American Institute of Electrical Engineers* 65 (12): 791–797.

13 Ansari, M.A., Martin, D., and Saha, T.K. (2019). Investigation of distributed moisture and temperature measurements in transformers using fiber optics sensors. *IEEE Transactions on Power Delivery* 34 (4): 1776–1784.

14 Presti, D.L., Massaroni, C., and Schena, E. (2018). Optical fiber gratings for humidity measurements: a review. *IEEE Sensors Journal* 18 (22): 9065–9074.

15 Zakaria, R., Mezher, M.H., Zahid, A.Z.G. et al. (2020). Nonlinear studies of graphene oxide and its application to moisture detection in transformer oil using D-shaped optical fibre. *Journal of Modern Optics* 67 (7): 619–627.

16 Yusoff, S.F.A.Z., Mezher, M.H., Amiri, I.S. et al. (2018). Detection of moisture content in transformer oil using platinum coated on D-shaped optical fiber. *Optical Fiber Technology* 45: 115–121.

17 Melcher, J., Deben, Y., and Arlt, G. (1989). Dielectric effects of moisture in polyimide. *IEEE Transactions on Electrical Insulation* 24 (1): 31–38.

18 A. Swanson, S. Janssens, D. Bogunovic, et al., "Real time monitoring of moisture content in transformer oil," in *Electricity Engineers Conference*, 2018.

19 Guo-Ming, M., Jun, J., Rui-Duo, M. et al. (2015). High sensitive FBG sensor for equivalent salt deposit density measurement. *IEEE Photonics Technology Letters* 27 (2): 177–180.

20 Wang, L., Fang, N., Huang, Z., and Abadie, M. (2012). *Polyimide-Coated Fiber Bragg Grating Sensors for Humidity Measurements*. InTech.

21 Sun, T., Grattan, K.T., Srinivasan, S. et al. (2011). Building stone condition monitoring using specially designed compensated optical fiber humidity sensors. *IEEE Sensors Journal* 12 (5): 1011–1017.

22 Alwis, L., Sun, T., and Grattan, K.T.V. (2013). Optical fibre-based sensor technology for humidity and moisture measurement: review of recent progress. *Measurement* 46 (10): 4052–4074.

23 Kung, P., Idsinga, R., -Durand, H.-C.V. et al. (2017). Fiber optics sensor monitoring moisture transport in oil in an operating transformer. In: *2017 IEEE Electrical Insulation Conference (EIC)*, 487–490. IEEE.

24 Laskar, S. and Bordoloi, S. (2013). Monitoring of moisture in transformer oil using optical fiber as sensor. *Journal of Photonics* 2013: 528478.

25 Razzaq, A., Zainuddin, H., Hanaffi, F., and Chyad, R.M. (2019). Transformer oil diagnostic by using an optical fibre system: a review. *IET Science, Measurement & Technology* 13 (5): 615–621.

26 Sequeira, F., Bilro, L., Rudnitskaya, A. et al. (2016). Optimization of an evanescent field sensor based on D-shaped plastic optical fiber for chemical and biochemical sensing. *Procedia Engineering* 168: 810–813.

27 Xu, Y., Fang, W., and Tong, L. (2017). Real-time control of micro/nano fiber waist diameter with ultrahigh accuracy and precision. *Optics Express* 25 (9): 10434–10440.

28 S. Yelikbayev, "Adiabatically tapered fiber-optic microsensor: Fabrication and characterization," Project Report, Nazarbayev University, Kazakhstan, 2020.

29 Wang, Y., Tan, B., Liu, S. et al. (2019). An optical fiber-waveguide-fiber platform for ppt level evanescent field-based sensing. *Sensors and Actuators B: Chemical*: 127548.

30 Brambilla, G., Xu, F., Horak, P. et al. (2009). Optical fiber nanowires and microwires: fabrication and applications. *Advances in Optics and Photonics* 1 (1): 107–161.

31 Zhang, L., Tang, Y., and Tong, L. (2020). Micro-/nanofiber optics: merging photonics and material science on nanoscale for advanced sensing technology. *iScience* 23 (1): 100810.

32 Wang, P., Wang, Y., and Tong, L. (2013). Functionalized polymer nanofibers: a versatile platform for manipulating light at the nanoscale. *Light: Science & Applications* vol. 2: e102.

33 Tong, L., Hu, L., Zhang, J. et al. (2003). Subwavelength-diameter silica wires for low-loss optical wave guiding. *Nature* 426 (6968): 816.

34 Tong, L., Hu, L., Zhang, J. et al. (2006). Photonic nanowires directly drawn from bulk glasses. *Optics Express* 14 (1): 82–87.

35 Holzki, M., Fouckhardt, H., and Klotzbücher, T. (2012). Evanescent-field fiber sensor for the water content in lubricating oils with sensitivity increase by dielectrophoresis. *Sensors and Actuators A: Physical* 184: 93–97.

36 T. V. Rao, V. Chakravarthy, and K. K. Murthy, "Working model of optical fiber sensor for estimation of sludge in oil in electrical transformer," vol. 49, no. 9, pp. 596–599, 2011.

37 Kim, T.-Y., Kim, J.-E., and Suh, K.S. (2008). On-line monitoring of transformer oil degradation based on fiber optic sensors. *Sensors and Materials* 20 (5): 201–209.

38 Onn, B.I., Arasu, P.T., Al-Qazwini, Y. et al. (2012). Fiber Bragg grating sensor for detecting ageing transformer oil. In: *2012 IEEE 3rd International Conference on Photonics*, 110–113. IEEE.

39 Mahanta, D.K. and Laskar, S. (2015). Oil-level measurement in power transformer using optical sensor. In: *2015 International Conference on Energy, Power and Environment: Towards Sustainable Growth (ICEPE)*, 1–4. IEEE.

40 Mahanta, D.K. and Laskar, S. (2015). Power transformer oil-level measurement using multiple fiber optic sensors. In: *2015 International Conference on Smart Sensors and Application (ICSSA)*, 102–105. IEEE.

41 Liao, J. (2011). Insulating oil level monitoring system for transformer based on liquid level sensor. In: *2011 Second International Conference on Digital Manufacturing & Automation*, 654–656. IEEE.

42 Antonio-Lopez, J.E., Sanchez-Mondragon, J.J., LiKamWa, P., and May-Arrioja, D.A. (2011). Fiber-optic sensor for liquid level measurement. *Optics Letters* 36 (17): 3425–3427.

43 Yeo, T.L., Sun, T., and Grattan, K.T.V. (2008). Fibre-optic sensor technologies for humidity and moisture measurement. *Sensors and Actuators A: Physical* 144 (2): 280–295.

44 Yulianti, I., Supa'at, A.S.M., Idrus, S.M., and Anwar, M.R.S. (2013). Design of fiber Bragg grating-based Fabry–Perot sensor for simultaneous measurement of humidity and temperature. *Optik* 124 (19): 3919–3923.

45 Alwis, L., Sun, T., and Grattan, K.T. (2012). Optimization of a long period grating distal probe for temperature and refractive index measurement. *Procedia Engineering* 47: 718–721.

4

Dissolved Gases Detection with Optical Methods

The failure of the transformer will cause supply interruptions and also will affect the utility electricity supply reliability. The cost of replacing a failed transformer is more than a million dollars. However, the failure of the transformer can be reduced by monitoring the insulating oil condition. In a transformer, oil acts as insulation, coolant, as well as the condition indicator. The electrical and thermal stresses in a transformer will generate fault gasses. The latter will be dissolved in the transformer oil and thus will reduce the insulation integrity. The amount of dissolved gas in the oil can be measured using the dissolve gas analysis (DGA) technique. The DGA is used to diagnose and detect incipient faults in the transformer based on the concentration of the dissolved gas contents in the transformer oil. The DGA is performed on the oil, which is extracted out from the transformer. Normally, the DGA test is carried out in the lab by using the gas chromatography (GC) analysis. The GC analysis has been used for the past 40 years and has shown success and gained trust among the electric utilities compared to other techniques. The main reason for the success is due to the sampling and analyzing procedures, which are cheaper and simple and are supported by many established standards related to the testing methods and analysis.

The fault gases (which mainly contain seven categories: H_2, C_2H_2, CH_4, CO, CO_2, C_2H_4, C_2H_6) can be seen as the symbol gases of the first available indication that dissolved inside the transformer oil; this process is known as the dissolved gases analysis technique. Therefore, DGA is considered to be the best approach and practice for evaluating a transformer's condition. Advantages of DGA include: advanced warning of developing faults; status checks on new and repaired units; convenient scheduling of repairs; and monitoring of units under potential overload conditions.

4.1 Online Dissolved Gases Analysis

The GC method is mostly used for quantitative measurement in the laboratory and default off-line routine tests. The traditional GC method has the advantages of high sensitivity, fast analysis speed, high separation efficiency, and low use of sample oils [1–13]. This chemical analysis method has become a regular project for the periodic detection of dissolved gas in power transformer oil. The method consists of several steps, from getting an oil sample to analyzing it in the laboratory. It includes extracting oil samples from the running transformer, using the degassing device to separate oil and gas, measuring the calibration of the gas chromatograph, and the completion of gas detection.

Optical Sensing in Power Transformers, First Edition. Jun Jiang and Guoming Ma.
© 2021 John Wiley & Sons Ltd. Published 2021 by John Wiley & Sons Ltd.

The weaknesses of the traditional GC method are listed as follows. GC systems are very sensitive to environmental conditions (temperature, humidity, atmospheric pressure, movement associated with vibration, or wind buffeting, etc.). While these conditions are perfectly controlled and checked in a laboratory, in the field this is much more challenging, with systems having to cope with day to night changes, winter to summer changes, weather patterns, and local shaking from transformer vibration, road traffic, and industrial processes. To provide a suitable environment for a field-based GC system is costly and complex, but this is the only way to guarantee long-term accurate results from a GC-based DGA system. Great care must be taken by the utility when assessing the guaranteed specification of operating conditions for the GC system in the field. This is often overlooked by the manufacturer in instrument specifications and only becomes obvious upon extended operation of the field GC. Its detection cycle is relatively long, sometimes even up to several months; thus it cannot track the equipment conditions in time and the tendency to change. In addition, the detection process is relatively complex, corresponding to too much cost in terms of time and money.

Though the GC method itself does have high precision, there might be a large error in the degassing process and the artificial correction of the gas absorption peak. Such a mistake would affect the overall test result.

Taking into consideration the above shortcomings and disadvantages of offline chromatographic detection methods, domestic and foreign scholars, relevant research institutions, and companies focus a lot on the research and development of transformer oil dissolved gas online monitoring technology and devices. As a supplement and development of the offline detection method of dissolved gas in oil, online monitoring technology has overcome a series of shortcomings, such as complex operation procedures, a long detection period, and large errors from man-made operation. Online monitoring technology has the advantage of high detection frequency, better real-time performance, and higher degrees of automation. In addition, the online monitoring technology uses the components and concentration of the gas dissolved in the oil as the basic characteristics. Establishing the online intelligent monitoring and fault diagnosis system of a transformer involves monitoring the operating conditions of the transformer, tracking the existence of potential failure, monitoring sudden failure, distinguishing and diagnosing the fault in real time, and making maintenance staff quicker to deal with the fault. Compared with the offline test method, which needs a lot of equipment and staff training and maintenance, the online oil dissolved gas monitoring technology performs much better in the aspect of economy, and hence improves the technical level of operation and maintenance a in transformer substation [12, 14–16]. A schematic of a possible valve installation location of an online DGA in a power transformer (not to scale) is illustrated in Figure 4.1 [2].

Various techniques for detection of dissolved gases in transformer oil have been proposed and implemented. With regard to online DGA detectors, there are mainly two categories: physical/chemical techniques and optical methods. Generally, physical and chemical gas detectors include thermal conductivity detectors (TCDs), flame ionization detectors (FIDs), semiconductor detectors, electrochemical detectors, catalytic combustion detectors, palladium gate FET (field effect transistor) detectors, array gas sensors, etc. The manufacturing processes of these detectors are relatively mature and the mass production costs are low. However, regular replacement and calibration will increase the workload of operation and maintenance due to the relatively poor

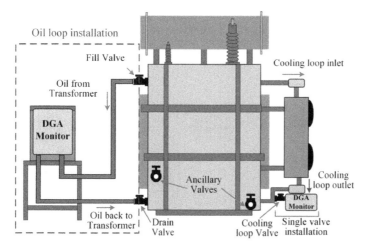

Figure 4.1 Installation schematic of an online DGA in a power transformer. Source: Modified from Bustamante et al. [2].

long-term stability of physical/chemical detectors. In addition, most physical/chemical techniques rely on the external carrier gases. In recent years, with the development of related technologies, solid oxide fuel cell sensors, carbon nanotube sensors, and nano metal-oxide semiconductor sensors have been proposed and researched. Part of the detector performance has been improved, but the problems, such as being susceptible to aging, saturation, electromagnetic compatibility, and impurity contamination, have to be considered in the future. Optical techniques came into the field of DGA owing to their superiority of non-contact measurement and immunity to electromagnetic interference.

4.1.1 General Quantitive Requirements of Online DGA

The relative distribution of the gases is therefore used to evaluate the origin of the gas production and the rate at which the gases are formed to assess the intensity and propagation of the gassing. Both kinds of information together provide the necessary basis for an evaluation of any fault and the necessary remedial action.

The online monitoring device for DGA is the hardware foundation for the analysis and diagnosis technique of dissolved gases in oil. The schematic is shown in Figure 4.2. The online monitoring device for dissolved gas in oil can determine whether there is an internal fault in the equipment, diagnose the fault type, and estimate the fault energy and other information of the fault point. There are two key points in the analysis technology of dissolved gas in oil: oil–gas separation technology and mixed gas detection technology.

However, the fault gases must be separated from the oil before measurements can be taken in existing techniques. Vacuum degassing, dynamic headspace degassing (as shown in Figure 4.3), oscillation degassing, membrane separation, etc., have been utilized for the oil–gas separation, and among them, the polymer membrane separation is frequently used because of its simple apparatus setup. Typically, the equilibrium time for membranes including polyimide (mainly for hydrogen analysis), PTFE (polytetrafluoroethylene), and fluorinated ethylene propylene (FEP, a.k.a. F46) require 24–72 hours,

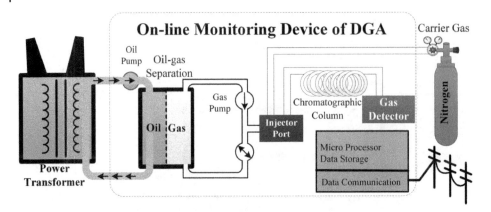

Figure 4.2 Illustration of an online monitoring device for oil-immersed power transformers.

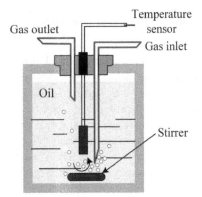

Figure 4.3 Typical headspace degassing/extraction unit.

so the separation process leads to a long response time and a complex structure; even worse, defects in gases cannot be located any longer. In this way, oil/gas separation isolates the characteristic transformer fault gases dissolved in the oil for spectral or chromatographic analysis. Therefore, the detection cycle is subjected to the efficiency of the oil–gas separation, which sometimes last for days [17–19].

This method of monitoring power transformers has been studied intensively and work is going on in international and national organizations such as CIGRE, IEC, and IEEE. Most of the DGA diagnostic tools in use today can be found in the IEEE C57.104 or IEC 60599 guides. Based on these two guides, other national and international guides that include additional tools are also available. The *IEEE Guide for the Interpretation of Gases Generated in Oil-Immersed Transformers* gives an extensive bibliography on gas evolution, detection, and interpretation. According to many practices, technical parameters for multi-gases online monitoring equipment are provided in reference [20].

After that, several recommended safe concentrations of individual gas values and total dissolved combustible gas (TDCG) have been summarized, from which the IEEE standards [21–23] are listed Table 4.1. A four-level criterion using data shown in Table 4.2 has been developed to classify risks to transformers.

The fault type can also be known using different diagnosing algorithms and recommended practices or experiences [24, 25]. According to IEC 60599, fault zone distributions can be seen in Figure 4.4 [26].

Table 4.1 Technical parameters for multi-gas online monitoring equipment.

Gases	Lower limit of detection (μL/L)	Upper limit of detection (μL/L)	Measurement errors requirement
H_2	2.0	2000	Lower limit of detection or ±30% (whichever is greater)
C_2H_2	0.5	1000	
CH_4	0.5	1000	
C_2H_6	0.5	1000	
C_2H_4	0.5	1000	
CO	25	5000	
CO_2	25	15 000	

Table 4.2 Dissolved gas concentrations at different status.

Status	Dissolved key gas concentration limits (μL/L)							
	H_2	CH_4	C_2H_2	C_2H_4	C_2H_6	CO	CO_2	TDCG
Standard	100	120	1	50	65	350	2500	720
Caution	101–700	121–400	2–9	51–100	66–100	351–570	2500–4000	721–1920
Warning	701–1800	401–1000	10–35	101–200	101–150	571–1400	4001–4010 000	1921–4630
Danger	>1800	>1000	>35	>200	>150	>1400	>10 000	>4630

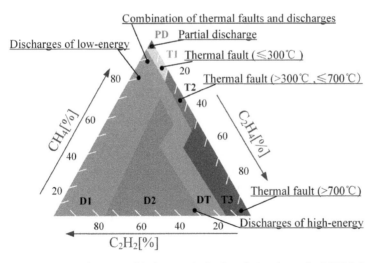

Figure 4.4 Coordinates and fault zones in the Duval triangle method (DTM). Source: Reprinted by permission of British Standards Institution (BSI) [26]. © 1999, BSI Standards.

4.1.2   Advantages of Optical Techniques in DGA

Domestic and foreign scholars, relevant research institutions, and companies have carried out a lot of work in the development and upgrading of online monitoring technology and devices for gas dissolved in the transformer oil. The achievements of previous research are mainly in the physical and chemical principles based on gas sensors. However, the feedback from its operation in the field is far from satisfactory, mainly due to its dependence on column gas separation, susceptibility to aging, saturation, impurity pollution, and other factors. Thus, the sensors need to be replaced or calibrated regularly, but the detection accuracy is difficult to improve.

In order to overcome the shortcomings of the various methods mentioned above, optical gas detection is introduced into the field of dissolved gas detection in transformer oil. This method has a series of advantages. It measures the quality and quantity of gas by processing spectral analysis or by optical measurement, thus being suitable for strong electromagnetic interference conditions. It can be achieved by non-gas separation of the column, so the detection cycle is short and the measurement efficiency is high. In addition, optical measurement and optical fiber communication do not consume the gas to be measured, which is helpful due to the need for repeat measurements of the same gas sample to reduce accidental error. Compared to physical and chemical detection technology, the performance of an optical device is reliable long-term monitoring and can be achieved without calibration or maintenance [15, 27–33].

In the following parts in this chapter, several optic-based techniques are illustrated and analyzed independently.

4.2   Photoacoustic Spectrum Technique

4.2.1   Detection Principle of PAS

Photoacoustic spectroscopy (PAS) is mainly based on the photoacoustic effect caused by gas molecules absorbing electromagnetic radiation (such as infrared). Different gases will absorb the specific wavelength of the infrared spectrum and the no radiation relaxation phenomenon makes the temperature of the gas rise and then release heat.

According to the principle of the PAS effect [10, 34–37], four procedures are mainly included, as illustrated in Figure 4.5 [38, 39]. The modulated light based on wavelength modulation or intensity modulation is beneficial to meet the requirement of an absorption band of the gases to be detected. Then selective absorption is excited to the higher energy state from the steady ground state after absorbing the energy from the light. However, the higher energy state is unstable and the molecule goes back to the ground state by colliding with another molecule. Translational energy is obtained during the colliding process, in which the gas pressure changes and the local temperature of the medium rises as well. Due to the modulated light source, the pressure periodically varies, resulting in the formation of thermal and acoustic signals in the target gas in the transformer oil sample. Usually, the signal is detected by an acoustic sensor and the acoustic signal is named the PA signal [40, 41].

By periodic modulation of the chopper chip, cyclical heat is generated and forms pressure waves, which produce an acoustic signal. The frequency of the pressure waves coincides with the frequency of the light source (chopper). The intensity of the sound

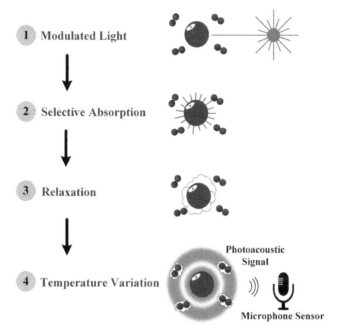

1 Modulated Light

2 Selective Absorption

3 Relaxation

4 Temperature Variation

Photoacoustic
Signal

Microphone Sensor

Figure 4.5 Principal illustration of photoacoustic spectroscopy detection. Sources: Ma [38] and Yang et al. [39].

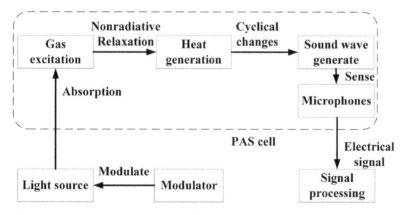

Figure 4.6 Principle of the photoacoustic spectroscopy technique.

wave can be detected by high sensitivity microphones. The gas concentration information can be obtained according to the proportional relationship between the intensity of the pressure wave and the concentration of the gas. The basic principle of the detection is shown in Figure 4.6.

Generally speaking, the target gas from the transformer oil absorbs specific light energy and converts it to a pressure wave (sound), which can be picked up by a microphone. The intensity of the sound is the concentration of the target gas and the spectrum, the sound intensity at different wavelengths, can be used to identify the absorbing components in the gases. In this way, dissolved gases in transformer oil are measured by PAS. To identify the traces of featured gases in insulation oil in practice, several procedures have been tried to improve the sensitivity. Firstly, highly sensitive microphones are used. The photo-acoustic detector is the key component in a PAS system and its high sensitivity helps to increase the lower detection limit of various gases. Secondly, a light source of high quality, especially large power, and tunable broadband laser, provides strong excitation and helps to reduce the noises and backgrounds. Thirdly, a specially designed chamber is necessary to increase the sensitivity and anti-vibration. Then last but not least, the specific strategy of the amplifier and filtering circuit topology is as a very powerful tool to improve the performance of the PAS measuring system. This allows for multiple gas species to be detected without the expense and complexity of multiple lasers while still achieving parts per million gas detection levels.

A diode laser or quantum cascade laser (QCL) is often used as a gas sample excitation source. The resolution of QCL-based spectral detection technology is higher than that of traditional spectral detection methods, and the detection does not require spectroscopic devices. Quartz-enhanced photoacoustic spectroscopy (QEPAS) is an improved version of the conventional microphone-based PAS and the topology of a standard QEPAS-based gas sensor system is shown in Figure 4.7 [38], where TA is the transimpedance amplifier; mR is the micro-resonator; ADM is the acoustic detection module; L is the plano-convex lens; and FC is the fiber collimator. In the QEPAS technique, a commercially available millimeter-sized piezoelectric element quartz tuning fork (QTF) is used as an acoustic wave transducer. By continuously tuning the laser wavelength, the absorption spectrum can be directly obtained from the photodetector. The detection sensitivity of the absorption spectrum is related to the optical power density of the laser light source, the effective absorption path, the system resolution, and the system noise. If the spectral power density of the QCL is high, the noise of the detector itself can be ignored, which is conducive to achieving high-sensitivity detection of gas. The line width of the laser line affects the system resolution and the resolution can be very small if the laser line is narrow enough.

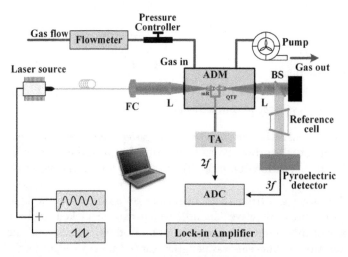

Figure 4.7 Schematic of the QEPAS measurement system. Source: Modified from Ma [38].

The QCL light source has good monochromaticity and a high degree of collimation, and a detection gas chamber with a long optical path can be used to greatly improve the effective absorption optical path of the gas in order to improve the gas detection sensitivity. With the merits of high sensitivity and selectivity, low cost, compactness, and a large dynamic range, QEPAS sensors have been applied widely in gas detection.

QCL can cover quite a wide wavelength range of a few micrometers to hundreds of micrometers, which meets the needs of absorption spectrum detection for a dissolved gas in transformer oil.

Moreover, QCL has the performance of engaging the frequency locking of the feedback light, and is convenient when combined with optical feedback cavity enhancement technology to enable high-precision and high-sensitivity gas detection.

An obvious enhancement of the QTF current signal can be obtained after adding two metallic tubes (mRs) to the QTF sensor architecture. To obtain a strong acoustic wave coupling, the length L of mR should be optimized.

4.2.2 Application of a PAS-Based Technique

There is no doubt that the device is vulnerable to noises for on-site applications. Those noise sources include: an electromagnetic and motor rotation noise in the field and microphone detection of the photoacoustic effect caused by the sound intensity is far less than the field noise; vibration of the cycle modulation chopper, the modulation frequency is low, and the excitation of the acoustic frequency is low; and spread to the pipe connected to the microphones. The typical structural illustration of PAS measurement can be seen in Figure 4.8 [10, 42].

These are just some of the issues associated with using what is essential for a laboratory technology. In 2011, GE Kelman Corporation developed Kelman TRANSFIX, a transformer oil in-line monitoring device (with a HITRON device detecting hydrogen). The system has a high detection sensitivity, wide detection range, no need of mixed gas separation, and a fast detection speed (up to once an hour); the typical composition is shown in Figure 4.9. It is a giant step for PAS gas detection to act as a possible alternative for field applications.

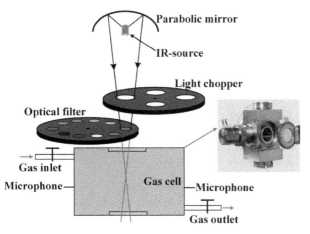

Figure 4.8 Structural illustration of photoacoustic spectroscopy detection. Sources: Bakar et al. [10]; Skelly [42].

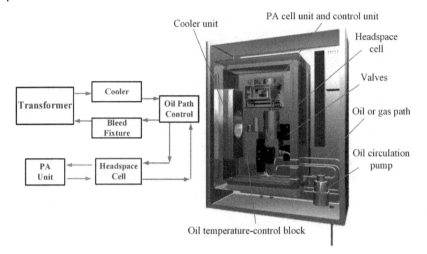

Figure 4.9 Typical structure and prototype apparatus based on PAS.

The PAS technology is a kind of indirect absorption spectroscopy from the perspective of a detection mechanism. The amplitude of the photoacoustic signal is proportional to the concentration of the gas and the size of the absorbed light energy, so it can reach an extremely high detection sensitivity with the development of laser light sources and the advancement of acoustic wave detector technology, especially in recent years. The key technology of a visible impact photoacoustic spectrometer to achieve high sensitivity detection of the target gas is a high-power laser light source and a high-performance acoustic wave detector.

Currently, the application of PAS for the detection of dissolved gases in transformer oil has some problems and challenges to be considered [2, 37, 43–45]. Firstly, the weak electrical signals of PAS are easily affected by the interference of the strong electromagnetic environment around the transformer, which seriously affects the stability and sensitivity of PAS detection. Secondly, interference signals such as vibration and noise around the transformer have a serious impact on the signal-to-noise ratio (SNR) of the photoacoustic spectrum detection. Thirdly, the power of the black body radiation source line is insufficient, the intensity of the excited photoacoustic signal is weak, and it is difficult for the minimum detection limit to be up to the industry standard requirements. Lastly, the optical power is prone to drift after long-term operation of the light source, which affects the measurement error of the PAS equipment.

4.2.3 Merits and Drawbacks

Compared to the conventional GC style technique and conventional methods, some advantages of using the PAS technique are very distinctive.

- **Less maintenance.** In the traditional chromatographic technology detection system, a carrier gas is required, and the chromatographic column is easy to be contaminated and needs to be replaced regularly. Normal atmospheric air can be used as the carrier medium, instead of high purity argon or helium or nitrogen, to reduce the replacement of carrier gases. As a further benefit, PAS-based detection does not require any account keeping about gas usage or logistics surrounding planning gas bottle replacement or delivering in a timely manner.

- **High sensitivity.** GC systems usually require frequent recalibration during operations relying on high purity gases. By contrast, PAS-based instruments produce a linear response to changing gas concentrations over a huge range up to approximately the ppm level with no need for multi-point recalibration in the field. The online detection system using PAS technology is technologically advanced and has high accuracy, which represents a new direction for the detection of faulty gases in oil in the future.
- **Competitive cost control.** The use of PAS technology to detect faulty gases in the oil can reduce maintenance costs, and the following investment is small, so it is a kind of cost-effectiveness. However, the system using chromatography technology requires a carrier gas and chromatographic columns that need to be replaced regularly, so the manual maintenance cost is high. In this sense, the photo-acoustic system is more competitive in terms of economy.

Although photo-acoustic technology has certain advantages for detecting faulty gases dissolved in oil, there are still some problems for further consideration.

- **Light source of high quality.** Multiple laser light sources or wide-range QCL laser sources are required for multi-component detection, so the system is more complicated and the cost increases. Since the photoacoustic signal is sensitive to factors such as temperature and pressure, it is necessary to consider how to eliminate the influence of temperature and pressure when designing a photoacoustic system.
- **Oil-gas separation dependent.** The photo-acoustic spectrum requires oil and gas separation, but no need for gas separation. It can be used to measure the mixed gas. The accuracy of the photoacoustic spectrum is related to the selection of the characteristic spectrum of the gas molecule to be measured, the sensitivity of the microphones, and the performance of the narrow-band filter. The PAS method nowadays has high sensitivity and can reach the ppm level. In addition, the required gas samples are less in the process.
- **Environmental influences.** When the PAS device is used in the transformer oil dissolved gas detection, it is sensitive to the oil vapor pollution and the noise source is widely existing. Those noise sources include: electromagnetic and motor rotation noise in the field, microphone detection of the photoacoustic effect caused by the sound intensity, which is far less than the field noise; vibration of the cycle modulation chopper, where the modulation frequency is low and the excitation of the acoustic frequency is low; and spread to the pipe connected to the microphones.
- **Restrictions on the components.** The wavelength resolution of the chopper usually used in the photoacoustic spectrum device is not fine enough, generally from the tens to hundreds in the nm range. The absorption spectrum of the near-infrared acetylene is given as an example. As shown in Figure 4.10, the distance between the two absorption lines of the acetylene gas is about 1.2 nm in the wavelength range of 1500 −1550 nm [46], each line width being about 0.045 nm, much smaller than the wavelength resolution of the chopper chip. Therefore, there is a certain degree of cross-sensitive problems of different gases. Although the separation algorithm and multimodal fitting can alleviate the cross-sensitive problem to a certain extent, there is no fundamental solution to the cross-sensitive problem.

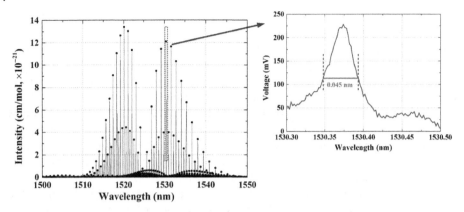

Figure 4.10 Absorption spectrum distribution of ethyne in wavelengths ranging from 1500 to 1550 nm.

All the reasons above (some of them are closely related) have limited the further development and popularization of the PAS technique.

4.3 Fourier Transform Infrared Spectroscopy (FTIR) Technique

4.3.1 Detection Principle of FTIR

When light passes through different components and different concentrations of gases, the light intensity can produce different degrees of attenuation. By measuring the intensity and attenuation of light, the change in gas concentration can be found.

When a certain volume of uniform gas is irradiated by a laser beam with a specific wavelength, the laser spectrum and the absorption spectrum of the irradiated gas overlap and the laser intensity is reduced, as shown in Figure 4.11.

The absorption of the gas in the light meets the Beer–Lambert law and the formula can be expressed as [47–50]

$$I(\lambda) = I_0(\lambda) \cdot \exp[-\alpha(\lambda) \cdot C \cdot L] \tag{4.1}$$

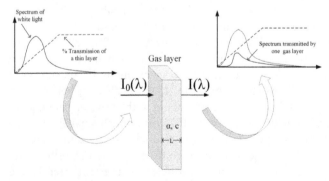

Figure 4.11 Principle of the Beer–Lambert law.

where

$I(\lambda)$ and $I_0(\lambda)$ are intensities of incident and emitted laser respectively, V.
C is the concentration of gas to be detected, $\mu L/L$.
L is the path length of the beam of light through the material sample, m.
$\alpha(\lambda)$ is the absorption coefficient per unit distance and per unit concentration of the gas, cm/mol.

The Beer–Lambert law in gas can also be written as

$$C = \frac{\ln[I(\lambda)/I_0(\lambda)]}{-\alpha(\lambda) \cdot L} \tag{4.2}$$

In Eq. (4.2), the concentration is easily obtained by finding the ratio between incident and emitted intensities.

Based on the Beer–Lambert law in infrared spectroscopy (IR), Fourier transform infrared spectroscopy (FTIR) is a technique used to obtain an infrared spectrum of emission or absorption, which can be available for solid, liquid, or gas [51–53]. The goal of FTIR is to measure how much light a sample absorbs at each wavelength. To evaluate the absorbance directly and accurately, a beam composed of many frequencies of light at once is excited for a sample and the action is repeated for different wavelengths and compositions to measure how much of the light is absorbed. With the repeated operations made in a short time span, some specific calculations are taken backward to estimate the absorption at each wavelength.

As an example, eight different dissolved gases in transformer oil have various wavelengths and their molecule absorption intensity distributions can be obtained from the HITRAN (high resolution transmission molecular absorption database, available calculated from website: http://hitran.org), as shown in Figures 4.12–4.19. The temperature is at 296 K and the cut intensity is set as 1E-28 cm/mol. It is easy to measure the specific concentration at the specific wavelength.

Since the Fourier transform spectroscopy device is widely developed and engaged in various applications, it is an easy way for it to be used in DGA.

Figure 4.12 Absorption spectrum distribution of H_2.

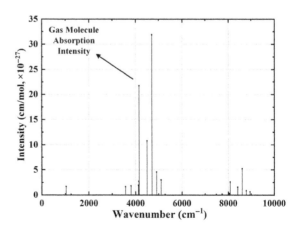

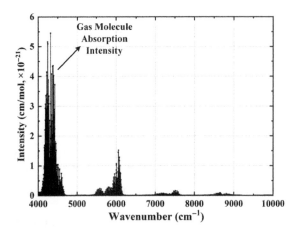

Figure 4.13 Absorption spectrum distribution of CH_4.

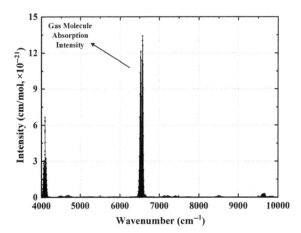

Figure 4.14 Absorption spectrum distribution of C_2H_2.

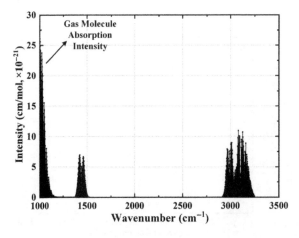

Figure 4.15 Absorption spectrum distribution of C_2H_4.

Figure 4.16 Absorption spectrum distribution of C_2H_6.

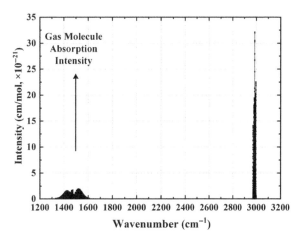

Figure 4.17 Absorption spectrum distribution of CO.

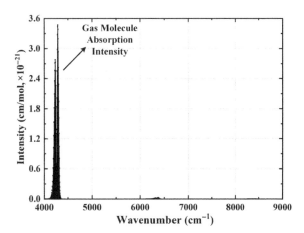

Figure 4.18 Absorption spectrum distribution of CO_2.

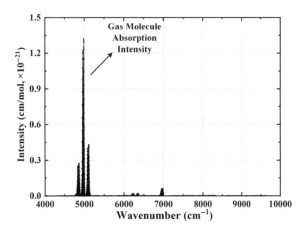

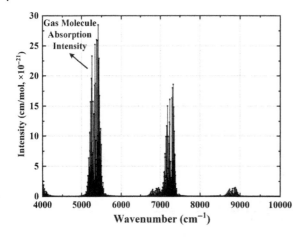

Figure 4.19 Absorption spectrum distribution of H_2O.

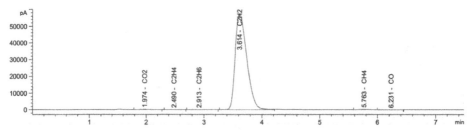

Figure 4.20 Typical result of gas chromatography. Source: Reprinted with permission from Jiang et al. [54]. © 2019 Elsevier.

4.3.2 Application of the FTIR-Based Techniques

4.3.2.1 FTIR Technique

Several oil samples having different feature gases can be prepared to be tested according to the IEC, ASTM standards. For example, the breakdown voltage is higher than 35 kV/mm, dielectric loss factor ≤ 0.005, moisture ≤ 10 μL/L, and the density @20 °C ≤ 859 kg/m^3. The typical result of conventional DGA is shown in Figure 4.20.

A developed or commercial FTIR device is used to obtain the specific infrared absorption spectrum of transformer oil, especially for acetylene as an example. Several oil samples with different acetylene concentrations have been prepared. E1, 89 660 μL/L; E2, 184 147 μL/L; E3, 202 814 μL/L; E4, 496 714 μL/L.

To bottle the oil sample, a quartz container with wide-band transmittance is installed in the FTIR device to verify the measurement. The illustration of the experimental setup can be seen in Figure 4.21.

To carry out the test, pure oil and oil with dissolved acetylene are involved to be compared and pure oil should be tested at the beginning. To ensure the accuracy of the test, it is recommended that every single sample is repeatedly tested 32 times or more, with the wave number resolution at 2 cm^{-1} or higher.

The issue of whether light can pass through the transformer oil and what kind of light can transmit through the oil needs to be addressed at the very beginning, since the spectral absorption characteristics are extremely complex with the composition of various chemical compounds.

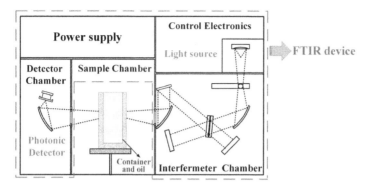

Figure 4.21 Experimental setup based on Fourier transform infrared spectroscopy. Source: Adapted with permission from Jiang et al. [54]. © 2019 Elsevier.

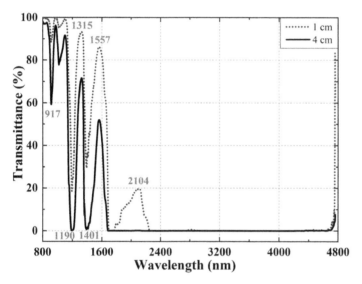

Figure 4.22 Transmittance distribution of pure oil in 1 cm and 4 cm optical path lengths. Source: Reprinted with permission from Jiang et al. [54]. © 2019 Elsevier.

During the test, the wavenumber ranges of $12\,500–2100\,\text{cm}^{-1}$ ($800–4762\,\text{nm}$) are selected to observe the transmittance distribution of infrared light, focusing on the light transmission of pure oil of both 1 cm and 4 cm optical path lengths, as shown in Figure 4.22. Among them, the mid-infrared absorption in the range of $4000–2100\,\text{cm}^{-1}$ ($2500–4762\,\text{nm}$) is very complicated. C—H bond, C—C single bond, carbon—carbon triple bond, etc., all have very strong absorption in this range, making the light transmittance on this section basically zero. Therefore, the relationship between absorbance A and transmittance T is $A = -\lg(T)$.

It is confirmed that the infrared transmittance range of transformer oil is very narrow, concentrated in the near-infrared wavelength range of 800–1687 and 1770–2261 nm. For the 1 cm optical path, the wavelength range where the light transmittance exceeds 50% includes 800–1158, 1244–1364, and 1472–1649 nm. As an example, the overall light transmittance at 1.557 μm of a 1 cm optical path cuvette is 85.6% and is 52.3% (just about to the fourth power of 85.6%) for the 4 cm optical path cuvette.

According to the test results, the light transmittance decreases with multiple power functions as the oil path of the transformer and the optical path of the cuvette increase. Therefore, in order to ensure that the infrared light source can penetrate the transformer oil, it is necessary to extend the oil path and optical path to balance in a reasonable range.

When the light path is longer, the mid- and far-infrared wavelengths are easily absorbed by the hydrocarbon groups in the oil, and the light transmission effect is obviously reduced. Therefore, the mid- and far-infrared wavelength range is not suitable for direct sensing of dissolved gases in transformer oil. In the process of subsequent sensor testing, the transmittance area is highlighted in the near-infrared band.

Since the peak value only reflects the spectral absorption intensity at a single and specific wavelength, and is easily affected by the baseline drifts, it is not the best characterized parameter of dissolved acetylene in the oil. On the contrary, spectral peak area (PA) is characterized by the absorption intensity within a period of wavelength interval. In order to obtain the peak area of the oil samples with different acetylene concentrations, the absorbance of each sample with a 1 cm path length is integrated. The principle of integration is to select the area of the absorbance curve above the baseline position. The test results of different acetylene oil samples are shown in Figure 4.23 and the gray shaded area is calculated as the peak area. From the test results, oil samples of different concentrations all have absorption peaks at 6518.16 cm^{-1} (1534 nm), but the baseline positions of different oil samples are different. This is because the Fourier Transform Infrared Spectrometer is a single-beam instrument, and the baseline drift is mainly caused by slight fluctuations in the energy of the light source and subtle changes in the environment. Baseline drift is a very normal phenomenon and can be improved by subtracting the measured background spectrum from each test.

The absorption intensity of acetylene in the gas phase and the oil phase (the blue line) are compared in the range of 1480 ~ 1640 nm, as shown in Figure 4.24.

The strongest absorption peaks of acetylene in the gas phase are around 6578.95 cm^{-1} (1520 nm) and around 6534.35 cm^{-1} (1530 nm) in the specific wavelength range. Compared with the absorption peak position in the gas phase, the central wavelength of the acetylene test in oil has a slight red shift phenomenon (the wavelength of the peak

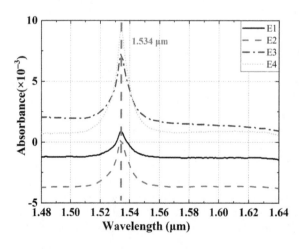

Figure 4.23 Absorbance distribution of oil samples in the range of 1480–1640 nm. Source: Reprinted with permission from Jiang et al. [54]. © 2019 Elsevier.

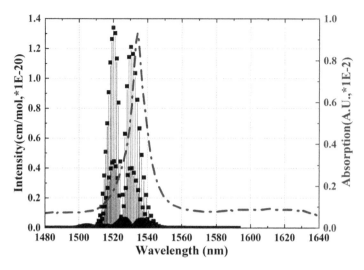

Figure 4.24 Absorption of acetylene in the gas phase and oil at wavelength of 1480–1640 nm. Source: Reprinted with permission from Jiang et al. [54]. © 2019 Elsevier.

absorption increases). In addition, the acetylene dissolved in oil presents an absorption spectrum with obvious width.

For the red shift of the position of the acetylene absorption peak in the oil, the distribution of the electron cloud is mainly considered. The electron cloud density of the triple bond of acetylene decreases due to the action of other molecules, and the strongest absorption peak is red-shifted. In addition, the distance between the acetylene molecules dissolved in transformer oil is much smaller than that in the gas phase, so it is easy to form a π–π conjugated system, which is connected by a single bond, and dislocation of π electrons occurs. When the total energy of the system decreases with the occurrence of the conjugation effect, then the wavelength red shifts.

In view of the broadening of the spectrum in oil, the transformer oil sample contains carbon—carbon bonds and carbon—hydrogen bonds, which are similar to acetylene gas. The distance between acetylene molecules in the liquid phase is much smaller than that in the gas phase, which reduces the stretching vibration. It exhibits a certain cumulative effect, which widens the spectrum width, and the corresponding full width at half maximum (FWHM) increases greatly. The force between gaseous molecules is small and negligible, and the molecules can rotate freely, so the vibrational rotation spectrum of the fine structure can be seen. While the molecular force is greatly increased in the liquid state, the rotation of the molecules becomes difficult. Therefore, only the superposition of the vibration rotation spectrum of each molecule can be detected, forming a broad spectrum. Moreover, the density of oil is much greater than that of acetylene gas, and the time interval between two adjacent collisions is less than the energy level lifetime. A continuous spectrum tends to be formed when the collision broadening is greater than the distance between adjacent spectral lines. Therefore, the broadening phenomenon of collisions is the main reason that the spectrum of dissolved acetylene in transformer oil is much larger than that of acetylene gas.

The wavelength range of 1504–1569 nm of the peak area integration covers the main absorption spectrum of acetylene gas molecules in the range of 1480–1640 nm, which

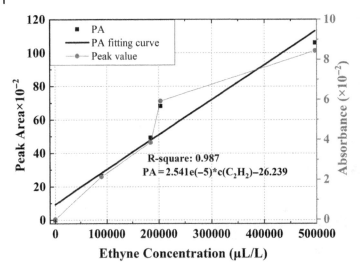

Figure 4.25 Fitting curves of peak area vs. acetylene concentration at a 1 cm optical path length. Source: Reprinted with permission from Jiang et al. [54]. © 2019 Elsevier.

shows the validity of the test and calculation data. The FWHM of the infrared absorption of acetylene gas is about 0.04 nm, so the absorption spectrum of acetylene dissolved in oil is extended by about 240 times.

The curves obtained by linear fitting the oil samples with different dissolved acetylene concentrations and the detected peak area are shown in Figure 4.25, in which PA represents the peak area and $c(C_2H_2)$ indicates acetylene concentration.

The linearity of the peak area fitting is only 0.987, but the peak area or peak value (all the effects of baseline drift are excluded) and the dissolved acetylene concentration in the oil sample are obviously positively correlated. Among them, the fitted value of the peak area reflects the change of the acetylene concentration better, mainly because the peak area characterizes the spectral absorption intensity in a certain wavelength range, while the peak can only reflect the spectral absorption intensity at a single wavelength with randomness and possible errors.

However, the measured sensitivity of dissolved acetylene is too low to satisfy actual needs. To improve the sensitivity, a particular quartz cuvette with an inner width of 4 cm is adopted in the follow-up experiment. Similarly, the final peak area and acetylene content data of different acetylene concentrations at a 4 cm optical path length are shown in Figure 4.26.

The higher the concentration of dissolved acetylene in the oil sample, the more obvious the corresponding absorbance. Compared with the test result of a 1 cm optical path, the slope of the absorption spectrum of an oil sample with a 4 cm optical path is larger. The reference line is selected within the peak area integration range to obtain the peak area value, and a linear fit is performed with the dissolved acetylene concentration in the oil, as depicted in Figure 4.27. The sensitivity of the 4 cm optical path achieves 1.020×10^{-4} (µL/L) $^{-1}$, which is approximately four times that of the 1 cm optical path length.

The SNR of the absorbance spectrum of the dissolved 972.5 µL/L acetylene oil sample is about 5, so the detection resolution of dissolved acetylene in the oil under the 4 cm

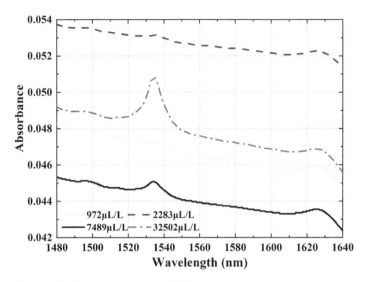

Figure 4.26 Absorption spectra of different acetylene oil samples at a 4 cm optical path.

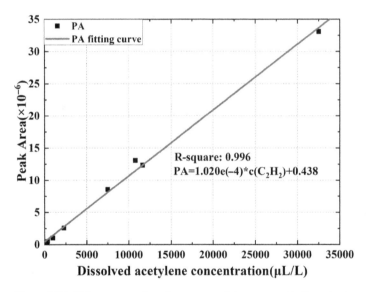

Figure 4.27 Fitting curves of peak area vs. acetylene concentration at a 4 cm optical path length.
Source: Reprinted with permission from Jiang et al. [54]. © 2019 Elsevier.

optical path reaches 194.5 µL/L, there are gaps to meet the requirements of the actual engineering detection. The feasibility of the FTIR technique for direct DGA detection is investigated.

4.3.2.2 Online FTIR Application
For the online analysis based on FTIR, the degassing unit is still preferred. An on-line analysis of oil-dissolved gas in power transformers using Fourier transform infrared spectrometry has been proposed and tested [55]. Several components are included and

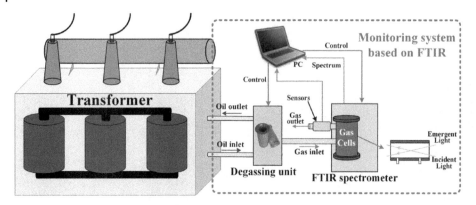

Figure 4.28 Schematic diagram of an oil-dissolved gas analyzer.

assembled into a cabinet like gas cells, degassing unit, data acquisition and control unit, temperature/pressure sensors, FTIR device, etc. It is vital to maintain the stable and comfortable temperature and pressure for the components in the cabinet and to calibrate the results. A possible diagram of the FTIR-based system is shown in Figure 4.28. The degassing unit, composed of the background cell and working cell, is used to extract oil-dissolved gas from the insulation oil of the power transformer. It is also important to decrease the volume of gas needed, which helps to increase the dynamics of the analysis system.

Another issue for the online application of the FTIR-based technique is on data processing. The problem is especially one of baseline drift or distortion in FTIR when analyzing gas continuously on site. To identify and treat spectral baseline drift, the common spectrum of standard gas analysis is obtained and corrected at first, absorbance of analytes is reduced from the corrected spectrum, and then Fourier transform needs to be performed with the difference spectrum. The first four-line strengths of the Fourier transform spectrum are proposed as feature variables to determine whether baseline distortion occurred [56]. The flowchart for identifying and correcting spectral baseline distortions during the process of gas analysis online with a Fourier transform infrared spectrometer is shown in Figure 4.29.

Most of the time, baseline distortion often shows itself as non-linear fluctuation. It is of great importance to increase the sensitivity in order to improve the algorithm.

4.3.2.3 Combination of FTIR and PAS

Based on the specific designed photo-acoustic cell gas detection system, which combines broad spectrum characteristics of Fourier infrared spectroscopy and the high sensitivity of photo-acoustic gas detection technology, it can achieve high-precision quantitative detection of dissolved gases in a variety of gases dissolved in transformer oil. It is expected to provide the foundation for the monitoring of operating state and fault-type analysis for the power transformer.

As an example, a high-precision T-type resonant photo-acoustic cell is proposed in [57] and an infrared PAS detection system is established with the Fourier transform infrared spectrometer, as depicted in Figure 4.30. CO_2 and C_2H_2 are selected as gas samples to carry out quantitative detection of dissolved gases in the insulation oil. The designed T-type resonant photo-acoustic cell is mainly composed of an absorption

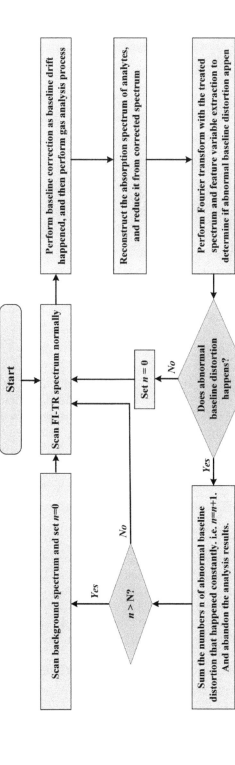

Figure 4.29 Identifying and correcting spectral baseline distortions for an online gas analysis.

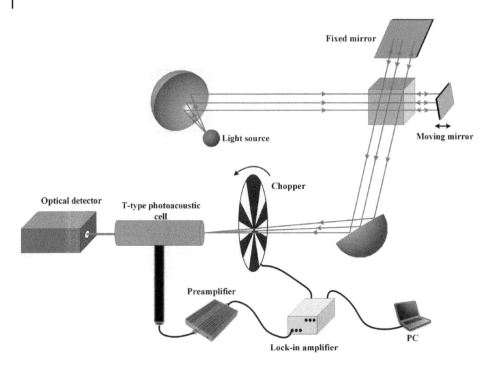

Figure 4.30 Schematic diagram of the FTIR-PAS system.

cavity and a resonant cavity that are perpendicular to each other. The acoustic detector is located at the top of the resonant cavity away from the incident light path to avoid the interference of noise caused by stray light from the photo-acoustic signal. The resonant frequency of the photo-acoustic cell is mainly determined by the resonant cavity. The resonant cavity is perpendicular to the incident light path and its length is not affected by the narrow space of the horizontal plane, so low-frequency resonance can be achieved in a limited amount to meet the sample space requirements of the spectrometer.

4.3.3 Merits and Drawbacks

It has been investigated whether it is possible to use the FTIR-based technique in the field of dissolved gases of a power transformer oil. Especially, different from conventional DGA measurements, there was also a discussion on the FTIR method as to whether it was even possible and feasible to get rid of the oil-gas separation process and bring in time-saving, simple, low cost, and quantitive measurements. A simple and direct detection of acetylene dissolved in power transformer oil through a spectral absorption method is available.

The FTIR is a mature module for stability and manufacturing, but there are some drawbacks being estimated.

- **Low sensitivity of direct measurement.** Dissolved gas can be detected through the oil directly with the access of infrared spectrum absorption, but the sensitivity is not high enough to meet the requirements, and the species of gases need to be further tested in the specific wavelength range. Moreover, consideration of oil temperature, coloration, and aging factors are not enough.

- **Vulnerable to ambient factors and surroundings.** The FTIR device itself is stable and more suitable for qualitative detection, but the quantitative analysis is vulnerable to temperature, pressure, vibration, etc., which is not capable for the field application.
- **More gases and cases to be verified.** A combination of FTIR and PAS is very helpful to ensure the sensitivity, but the category of gases is still limited to be qualified for continuous monitoring health status of insulation oil in power transformers in the field.

4.4 TDLAS-Based Technique

4.4.1 Detection Principle of TDLAS

Hydrocarbon gas is an important part of the dissolved gas-in-oil, mainly including methane, acetylene, ethane, and ethylene four hydrocarbon small molecule gases. All of them have infrared absorption characteristics and can be detected by infrared absorption spectroscopy. However, how the very weak absorption signals on the background can be measured to overcome the influence of disturbances is the challenge in achieving absorption detection based on the infrared spectrum.

The tunable diode laser absorption spectrum (TDLAS) technique is an alternative of direct absorption spectrum by measuring a high-resolution target spectral region with a tunable sweeping diode laser source [49, 50, 52, 58–61]. The TDLAS line scan may either be a straightforward linear sweep of the wavelength (direct absorption) or include an additional modulation on top of the wavelength sweep (wavelength modulation spectroscopy, WMS). The latter approach gives a better SNR and clearer identification of the baseline beneath an absorption line, as shown in Figure 4.31 [62]. Different gas molecules have their own absorption line frequency and line type. Based on the sensing of the absorption line of the gas molecule to be measured, the tunable laser spectral absorption tropology can be used to trace specific and unique gases.

When there is no gas of interest absorbed in the light path, PD output is proportional to the modulated intensity of the tunable laser source, as shown in Figure 4.32a and b. Once the gas of interest exists, the absorption region appears during the PD detection cycle, and the absorption signal is easy to obtain through a lock-in amplifier, as shown in Figure 4.32c and d. Since the output intensity of the laser is varied with the injected modulation current, the 2f harmonic signal in Figure 4.32e is actually asymmetric around the central position [62].

It is possible to establish the relationship between the gas concentration and the harmonic signal. Due to the amplitude of a high order of the harmonic component, the second harmonic signal (2f) is preferred to be detected and calculated.

The output of the laser diode (LD) is controlled by the value of the injection current. To modulate the absorption spectrum, a cosine current is applied to the injection current i_{ic} to drive the LD:

$$i_{ic} = i_{cen} + i_{m} \cos \omega_{m} t \tag{4.3}$$

where i_{cen} and i_{m} are the central current value and amplitude of the cosine current and ω_{m} is the modulated frequency. Then, the instantaneous frequency output of the LD can

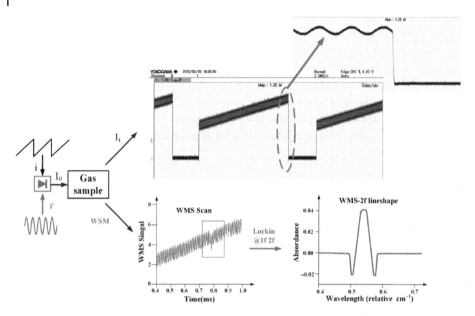

Figure 4.31 Illustration of gas concentration measurements by a second harmonic wave. Source: Reprinted with permission from Jiang et al. [62]. © 2018 Institute of Electrical and Electronics Engineers.

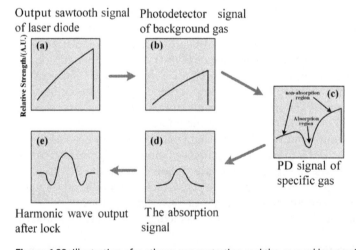

Figure 4.32 Illustration of methane concentration and the second harmonic wave signal.

be expressed as

$$v(t) = v_{cen} + v_m \cos \omega_m t \tag{4.4}$$

where v_{cen} and v_m are the central frequency value and amplitude of the wavelength. The intensity of the emitted laser is given by

$$I[v(t)] = I_0[v(t)] \cdot \exp\{-\alpha[v(t)] \cdot c \cdot L\}. \tag{4.5}$$

In the application of tracing gas detection, it is can be seen that $\alpha[v(t)] \cdot c \cdot L \le 1$. Therefore, the intensity of the emitted laser is simplified as

$$I[v(t)] = I_0[v(t)] \cdot \{1 - \alpha[v(t)] \cdot c \cdot L\}. \tag{4.6}$$

In an ideal condition, suppose that the output intensity of the laser is kept stable independently of the output wavelength. Then

$$I_0[v(t)] = I_0(v_0) = I_0. \tag{4.7}$$

Set $\theta = \omega_m t$ and Eq. (4.7) can be written as

$$I(v_{cen}, \theta) = I_0 \sum_{n=0}^{\infty} A_n(v_{cen}) \cdot \cos(n\theta) \tag{4.8}$$

where $A_n(v_{cen})$ is the harmonic component and n refers to the order of the harmonic component:

$$A_n(v_{cen}) = \frac{2c \cdot L}{\pi} \int_0^{\pi} -\alpha(v_{cen} + v_m \cos\theta) \cos(n\theta) d\theta \tag{4.9}$$

where the harmonic component can be obtained by a lock-in amplifier. It is easy to find that there is a certain relationship between the gas concentration to be detected and the harmonic component.

At the point of v_{cen}, the Taylor series of absorption coefficient is calculated as

$$A_n(v_{cen}) = \frac{2^{1-n} c \cdot L}{n!} v_m^n \frac{d^n \alpha}{dv^n}\Big|_{v=v_{cen}}. \tag{4.10}$$

The linear relationship of the harmonic component and the absorption coefficient is confirmed. The absorption coefficient of an odd order harmonic component is zero at the central position, while an even order is the maximum. Since the amplitude of the harmonic signal decreases with increasing order, in practice the second harmonic component is selected as the detection signal. Through the peak–peak value of the second harmonic signal from a lock-in amplifier, the concentration information of the to-be-detected gas can be inverted:

$$A_2(v_{cen}) = \frac{I_0 \cdot c \cdot L}{4} v_m^2 \frac{d^2 \alpha}{dv^2}\Big|_{v=v_{cen}}. \tag{4.11}$$

However, the residual amplitude modulation (RAM) phenomenon [49, 50, 63, 64] occurs when the harmonic decomposition is performed in a TDLAS system due to the adoption of the WMS. The second harmonic signal is not symmetrical about the center frequency point, so it is necessary to select the peak-to-peak voltage value for the calculation when using the second harmonic signal to invert the concentration of the gas to be measured.

Rather than the relative strength of the absorption signal with WMS, there is no need to determine the background line during the measurement process, and a higher sensitivity measurement can be achieved. According to the above analysis, compared with the direct absorption method, the tunable laser spectroscopy method can significantly improve the detection sensitivity.

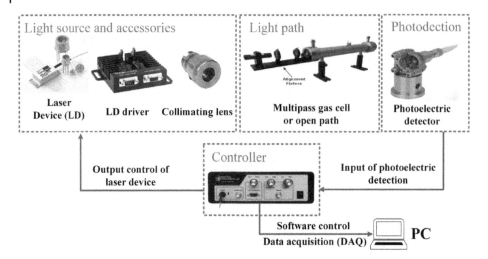

Figure 4.33 Main hardware components in a typical TDLAS system. Source: Reprinted with permission from Jiang et al. [65].

4.4.2 Application of the TDLAS-Based Technique

Generally, a typical TDLAS hardware system is mainly composed of a light source and its drive unit, gas cell (absorption light path), photodetector, control, and DAQ device, as shown in Figure 4.33 [65]. It is relatively mature to utilize the TDLAS system to detect industrial gases in various applications. However, multiple gases measurement, cross-interference, high sensitivity, and vibration of practical applications should be taken into consideration, but it is not possible to apply a similar device to online dissolved gases detection in a power transformer directly.

For a tunable semiconductor laser absorption spectroscopy technique on the basis of wavelength demodulation and the harmonics amplitude to trace gas concentrations, there are mainly three aspects to guarantee and improve the detection, as shown in Figure 4.34.

1. High-performance, narrow linewidth laser light sources are selected to increase the absorption coefficient of every single-component gas detection and to avoid cross-interference of several gases.
2. An integrated long optical path gas cell is specially designed and processed to increase the effective absorption optical path, enhance the anti-vibration performance, and ensure the stability of the output. Moreover, high reflectivity of the mirror surface is helpful to avoid the superposition of etalon stripes and reduce system noise in order to improve the sensitivity.
3. Multi-gas component measurement topology with time-sharing switching is considered to reduce optical path loss. Each light source output and temperature control are independently controlled to avoid cross-interference. Circuit isolation is ensured through time-sharing control of each gas detection unit.

In addition, the influence of temperature and pressure on gas sensing is clarified and calibrated through tests. Then environmental parameters need to be determined and the modulation factor and detection sensitivity are improved accordingly.

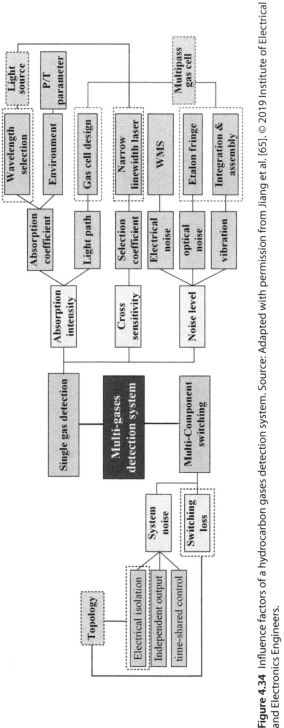

Figure 4.34 Influence factors of a hydrocarbon gases detection system. Source: Adapted with permission from Jiang et al. [65]. © 2019 Institute of Electrical and Electronics Engineers.

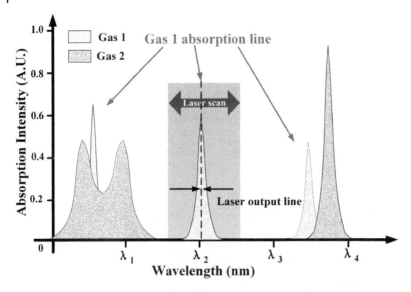

Figure 4.35 Schematic diagram of the absorption line selection. Source: Adapted with permission from Jiang et al. [65]. © 2019 Institute of Electrical and Electronics Engineers.

4.4.2.1 Optical Lasers

In the TDLAS system, the laser is not only a light source but is also a guarantee of spectral subdivision. In the selection of the absorption spectrum, two factors are emphasized: one is the intensity of the absorption spectrum, the other is to avoid interferences between the spectrum lines including H_2O, CO, CO_2, CH_4, C_2H_2, C_2H_4, and C_2H_6. The diagram of the absorption line selection can be found in Figure 4.35.

Actually, basic gas absorption bands are generally located in the mid-infrared region (MIR). High detection sensitivity can be obtained using mid-infrared lasers. However, different gases have different absorption peaks, making the system complicated, expensive, and difficult to integrate. Near-infrared region (NIR) lasers can operate at room temperature and can cover various absorption bands, but the gas absorption intensity is weak. Long-path absorption cells and WMS techniques can be used to increase detection sensitivity. Therefore, the near-infrared region (defined by the ASTM Working Group on NIR as 780–2526 nm) is preferred in this research. Last but not least, the transmission loss of the fiber is relatively low in NIR, especially optical telecom bands (O band: 1260–1360 nm; E band: 1360–1460 nm; S band:1460–1530 nm; C band: 1530–1565 nm; L band: 1565–1625 nm; U band: 1625–1675 nm), as shown in Figure 4.36. With the aforementioned consideration, this study focuses on absorption lines in telecom bands.

From the HITRAN and Pacific Northwest National Laboratory (PNNL) database, the absorption line information in a specific range can be searched and queried. Different hydrocarbon gases are determined and selected in the near-infrared range, as shown in Table 4.3.

Taking methane gas as an example, the absorption peak of methane is selected at 1653.72 nm. From the absorption spectrum, other gases have no obvious absorption peaks at this wavelength. In principle, interference from other gases can be effectively avoided.

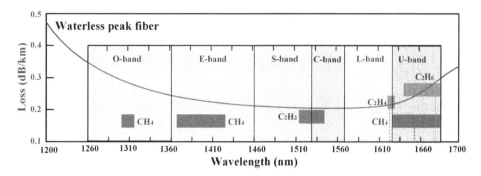

Figure 4.36 Absorption wavelength distribution of four hydrocarbon gases. Source: Adapted with permission from Jiang et al. [65]. © 2019 Institute of Electrical and Electronics Engineers.

Table 4.3 Central wavelength of C_2H_2, CH_4, C_2H_4, and C_2H_6.

Gases	Typical wavelengths with high intensity (nm)	Recommended wavelength (nm)
C_2H_2	1310, 1531, 1533	1530.37
CH_4	1647, 1651, 1654, 1660	1653.72
C_2H_4	1620, 1627	1620.04
C_2H_6	1647, 1651, 1680	1679.06

4.4.2.2 Multi-pass Gas Cell

According to the Beer–Lambert law, the optical path of the TDLAS system has a great impact on the system performance to trace gas detection. For certain gases, the signal attenuation caused by gas absorption is tightly related to the concentration of the gas and the optical path length. The intensity of the attenuated signal can be effectively increased with a long optical path. It is important to extend the optical path to achieve high sensitivity with multiple passes, including typical White and Herriott structures.

Since a Herriott cell has a very simple composition with only two spherical mirrors, it is easier to adjust the optical path and so the design is more suitable for application in a power transformer. For example, a customized Herriott gas cell with a physical length of 0.3 m reaches a multi-pass effective optical path of about 10.2 m long. The multi-pass reflection spot patterns of far-end and near-end mirrors are depicted in Figure 4.37.

In addition, as the spectral width and amplitude of the to-be-measured gas are related to the ambient temperature and gas pressure in the cell, it is necessary to design a specialized Herriott cell that considers temperature, pressure, and vibration.

A verified multi-pass gas cell has been designed with a temperature sensor, pressure sensor, fixed photodetector, and auxiliary components, as shown in Figure 4.38. Details of the parameters of the gas cell are shown in Table 4.4. The specified gas cell has advantages of small size, a long optical path, strong absorbance, and high sensitivity.

In special cases, a heating belt is wrapped around the cell to maintain a stable temperature for the field application.

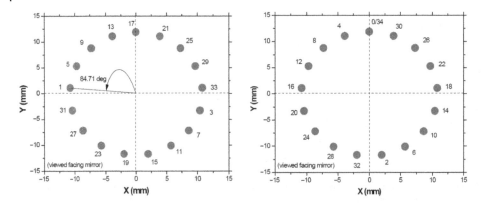

Figure 4.37 Reflection spot pattern of a far-end and near-end mirrors. Source: Reprinted with permission from Jiang et al. [62]. © 2018 Institute of Electrical and Electronics Engineers.

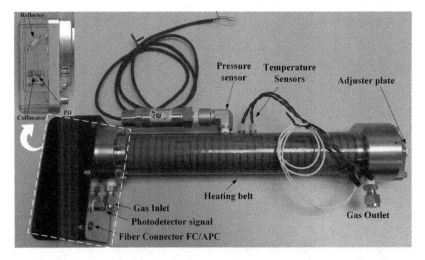

Figure 4.38 Structure view of the specialized long-path multi-pass cell. Source: Reprinted with permission from Jiang et al. [65].

4.4.2.3 Topology of Multi-gas Detection

In the practical application of TDLAS, near-infrared lasers are preferred due to the low price of light sources and the corresponding photodetectors, although the absorption intensity of gas is relatively weak. The challenge is that different central wavelengths of the gases need various laser sources individually, together with their corresponding thermoelectric controller (TEC) modules.

Take the four hydrocarbon gases as an example. Four sets of lasers and the drivers are needed to control four incident light and one output light. Thus, an optical coupler or a fiber switch is essential to switch the laser channels in sequence to the Herriott cell. The results of two optical components are compared in the same condition, as shown in Figure 4.39. The fiber switch is selected since the optical switch scheme has a much better SNR performance.

Once the optical switch is determined, the configuration of the TDLAS-based hydrocarbon gas detection system can be checked, as depicted in Figure 4.40. The entire

Table 4.4 Technical parameters of the specialized Herriott cell.

Items	Specifications
Optical path length	10.13 m
Volume of gas cell	0.24 L
Number of reflections	34
Reflectance of reflector	≥98.5%
Angle of incidence	2.07°
Diameter of coupling hole	3.3 mm
Maximum beam diameter	3.0 mm
Surface accuracy of mirror	1/10 Wavelength
Surface roughness of mirror	20 ~ 10
Mirror coating	Gold (HfO_2)
Length of gas cell	340.0 mm

Source: Reprinted with permission from Jiang et al. [65]. © 2019
Institute of Electrical and Electronics Engineers.

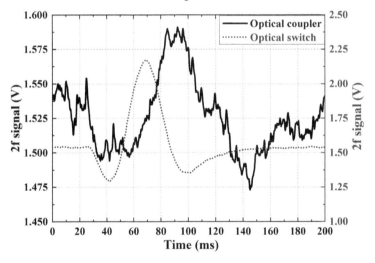

Figure 4.39 Comparison of optical coupler vs. optical switch. Source: Reprinted with permission from Jiang et al. [65]. © 2019 Institute of Electrical and Electronics Engineers.

system consists of four parts: the light source control unit, the laser launch and receive unit, the gas cell and gas path, and the data acquisition unit. The light source control unit comprises four FPGAs, four TECs, four DFB lasers, and an optical switch.

It should be mentioned that the operation time of the TDLAS detection system includes: oil injection, less than 1 minute; vacuum pumping, less than 1 minute; oil-gas separation, 2–10 seconds; gas detection and data processing, less than 10 seconds; others (for example, open/close valve), less than 1 minute. Therefore, the whole detection cycle is less than 5 minutes and the real-time measurement is quite beneficial to online monitoring and diagnosis, as shown in Figure 4.41.

Taking single gas detection as an example, the trace gas concentration detection process of the TDLAS system is as follows: the trace gas dissolved in the transformer oil to

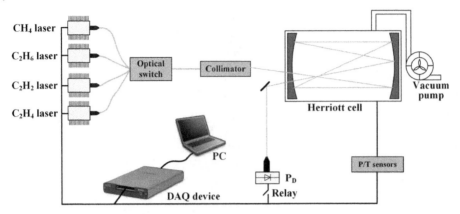

Figure 4.40 System configuration of TDLAS for hydrocarbon gases in transformer oil.

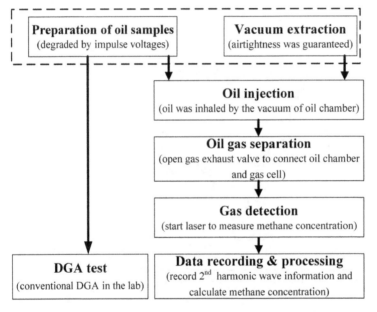

Figure 4.41 Procedures of TDLAS measurement. Source: Reprinted with permission from Jiang et al. [47]. © 2016 Institute of Electrical and Electronics Engineers.

be detected is passed into the long optical path absorption cell. The driver adjusts the current and temperature to ensure the operation of the semiconductor laser. At the same time, it realizes the tuning output of the semiconductor laser according to the modulation signal given by the modulation signal source, and then enters the long optical path absorption cell after passing through the collimator. After being fully absorbed, the output light enters the photodetector. The optical signal containing the absorption information of the to-be-measured gas is converted into an electrical signal. The electrical signal is amplified by the primary gain of the pre-amplifier and then the lock-in amplifier. With the harmonic signal acquired after analog-to-digital conversion, subsequent processing is done to transfer the gas concentration.

4.4.2.4 Laboratory Tests

In order to detect the sensitivity of hydrocarbon gas sensing, measurements of spectral absorption of methane, acetylene, ethylene, and ethane are completed after the oil–gas separation. Certain concentrations of hydrocarbon gas go through the specialized multi-pass cell with the different flowrates of the balance gas (N_2) versus the target gases. Typically, the 2f signal of the absorption spectrum (CH_4 as an example) is shown in Figure 4.42. Due to the RAM phenomenon, the peak-to-peak voltage value (PPV) of the 2f signal is preferred as the amplitude. With different concentrations, the relationship of the specific gas and the PPV can be established.

For example, the linear curving for methane detection is obtained in Figure 4.43 in specific circumstances and design. The unit of each gas concentration is µL/L. The fitting curves indicate that the sensor has good linearity and good sensitivity. In this situation, the concentration of methane increases by 100 µL/L and the voltage increases by 0.086 V.

Similarly, PPV of the 2f signal of ethyne with different concentrations is shown in Figure 4.44. To investigate the detection performance of the various gases, it is essential that the noise level and the SNR are considered. As to the gas detection, the noise amplitude is controlled at the mV level. Accordingly, the detection sensitivity of methane and ethyne reaches 1.2 (µL/L)/mV and 0.4 (µL/L)/mV, respectively.

Judging from the detection results, the developed hydrocarbon gas detector can realize the sensing of four hydrocarbon gases, especially the detection of typical characteristic gases (methane and acetylene) at the level of ppm or even sub-ppm.

Since various gases have multiple absorption peaks, the absorption peaks of different gases may overlap. Thus, it is also necessary to carry out cross-interference tests to avoid possible interferences of the dissolved gases in the oil.

In order to evaluate the influence of temperature and pressure on the results, it is necessary to carry out tests with different temperature and pressure tests under certain

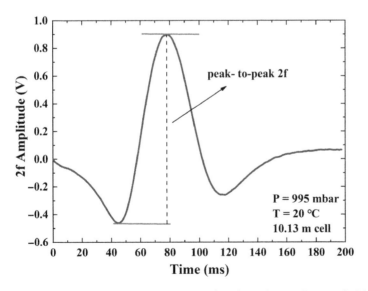

Figure 4.42 Typical absorption spectrum of methane. Source: Reprinted with permission from Jiang et al. [65]. © 2019 Institute of Electrical and Electronics Engineers.

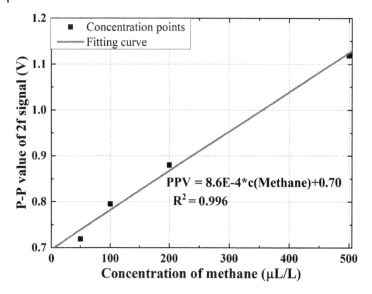

Figure 4.43 2f signals of methane at different concentrations. Source: Reprinted with permission from Jiang et al. [65]. © 2019 Institute of Electrical and Electronics Engineers.

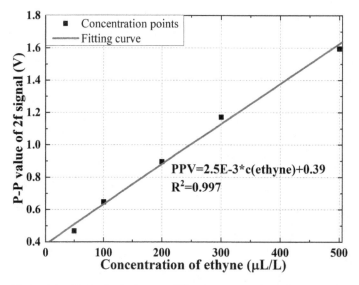

Figure 4.44 2f signals of ethyne at different concentrations. Source: Reprinted with permission from Jiang et al. [65]. © 2019 Institute of Electrical and Electronics Engineers.

concentrations in order to provide a basis for the final determination of temperature and pressure for the setup.

Taking methane as an example, an experiment is recorded with the ambient temperature of 25.4 °C and relative humidity (RH) of 82%, and TDLAS-based second harmonic signals at different temperatures and pressures, as shown in Figures 4.45 and 4.46. In

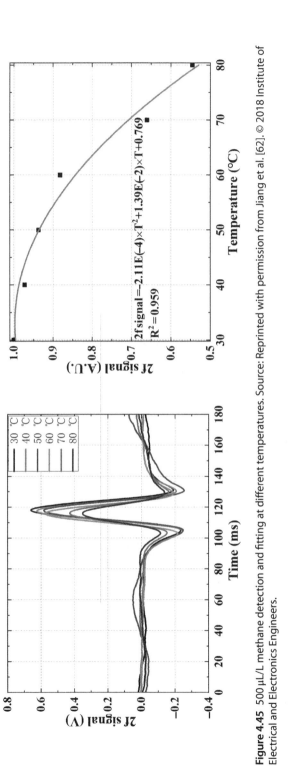

Figure 4.45 500 µL/L methane detection and fitting at different temperatures. Source: Reprinted with permission from Jiang et al. [62]. © 2018 Institute of Electrical and Electronics Engineers.

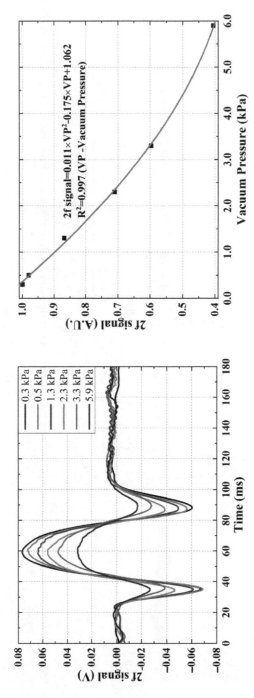

Figure 4.46 500 µL/L methane detection and fitting at different pressures. Source: Reprinted with permission from Jiang et al. [62]. © 2018 Institute of Electrical and Electronics Engineers.

this case, the 2f peak-to-peak value is normalized and the 2f signal is normalized as an arbitrary unit (AU).

During the heating process, the pressure in the gas path increases slightly, which has so little effect on the test results that it can be ignored. With the increase in temperature, the detected peak value gradually decreases. The amplitude attenuates by about 12% after an increase of 30 °C and by about 45% after an increase of 50 °C. Then the peak-to-peak values of the harmonic signal are normalized and fitted; the second-order curve fits well, as shown in Figure 4.45.

Similar to the temperature test, the absorption spectrum obtained under different pressures is shown in Figure 4.46, where 101.3 kPa is standard atmosphere (1 atm). As the air pressure increases, the peak-to-peak value of the second harmonic signal shows a significant downward trend.

Environmental parameters and influences of temperature and vacuum pressure have been investigated in the laboratory. With the above tests and data, several points should be considered.

1. As the temperature increases, the absorption intensity gradually weakens. In addition, the FWHM also shows a trend of widening. The main reason is that the increase in temperature makes the kinetic energy of the gas molecules increase, which makes it easier to be excited, consumes less photon energy, and weakens the absorption intensity.
2. In the temperature range of 30–50 °C, the temperature has little influence on the amplitude of the second harmonic. While the temperature range of the industrial site is generally located at −20–50 °C, it is possible to construct a gas chamber with a constant and appropriate temperature control module. The temperature setting can be close to the application requirement to avoid temperature interference to the results of the TDLAS system.
3. As the pressure decreases, the amplitude of the detected harmonic signal increases slightly. In addition, low air pressure is beneficial to improve the efficiency of oil/gas separation, so low air pressure is preferred. However, very low air pressure places very high requirements on the airtightness of the system and the performance of the vacuum pump, and at the same time it extends the vacuuming time and detection cycle. Therefore, it is reasonable to set the detection air pressure to about 1 kPa.

4.4.2.5 Field Application

Online DGA based on the proposed optical sensing technique was deployed in the field. Transformer oil flows out from the oil outlet valve located at the bottom of the transformer. The oil outlet valve flows to the detection device through copper pipe, as illustrated in Figure 4.47. The oil outlet valve is located at the lower part of the transformer, where the oil pressure is relatively high, which is conducive to the flow and circulation of transformer oil.

The system was installed in a substation, integrated with a TDLAS-based hydrocarbon gases sensing system. Online detection of transformer dissolved gases using the TDLAS technique-based equipment [62], as shown in Figure 4.48, has been conducted. A comparison test between a developed online DGA device and existing DGA equipment was carried out in the same circulating oil path of the main transformer.

After commissioning the on-site oil circuit circulation and gas detection device, the initialization of the oil sample is completed. The field test results of the TDLAS device

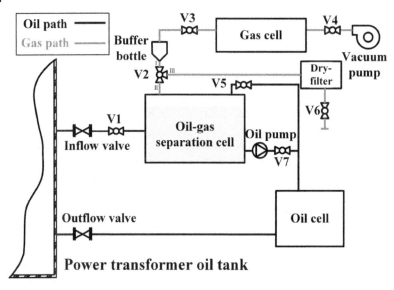

Figure 4.47 Mechanical structure of a multi-gas detection system in the field. Source: Reprinted with permission from Jiang et al. [65]. © 2019 Institute of Electrical and Electronics Engineers.

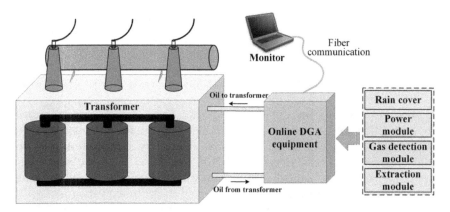

Figure 4.48 Typical online DGA equipment in the field based on the TDLAS technique. Source: Based on Jiang et al. [62].

are compared with the conventional DGA equipment of the transformer. After repetitive measurements, the comparative test data, especially four hydrocarbon gases and the total hydrocarbons (THC) values, are shown in Table 4.5.

The high consistency of the field test data and the high agreement of the TDLAS results to that of the reference DGA manifests the practicality of the TDLAS-based DGA device in monitoring symbol gas concentrations of transformer oil in the field.

Moreover, a contrasting analysis is provided in Table 4.6, indicating that the TDLAS system is capable of achieving the commission of online DGA monitoring according to industry requirements.

In summary, this is an alternative approach using tunable diode laser absorption spectroscopy to realize multiple gases detection of DGA for the health status of power transformer oil with consideration of a special design in the light source, multi-pass gas cell,

Table 4.5 Comparison results of developed equipment vs. an existing system on site.

Tests	$CH_4(\mu L/L)$	$C_2H_6(\mu L/L)$	$C_2H_4(\mu L/L)$	$C_2H_2(\mu L/L)$	THC($\mu L/L$)
1	0.551	1.470	1.887	0.093	4.000
2	0.384	1.070	1.398	0.022	2.874
3	0.555	1.407	1.508	0.059	3.529
Average	0.497	1.316	1.598	0.058	3.468
Reference DGA	2.03	0.75	1.53	0.00	4.31

Source: Adapted with permission from Jiang et al. [47, 65].

Table 4.6 Contrasting analysis of gas concentration detected by the TDLAS equipment.

Items	H_2 ($\mu L/L$)	C_2H_2 ($\mu L/L$)	THC ($\mu L/L$)
TDLAS equipment	3.542	0.058	3.468
Reference DGA system	1.10	0.000	4.310
Error value	2.442	0.058	0.842
Error limit	5.0	1.0	–
Alert value	150	5	150

topology, etc. In addition, environmental calibration helped the field tests and enabled the effectiveness of using the TDLAS approach to diagnose health conditions for power transformers.

4.4.3 Merits and Drawbacks

TDLAS is a new approach for dissolved gas measurement for power transformer oil. The merits can be categorized as follows:

- **High sensitivity.** In nature, the tunable laser offers a precise approach for flexible control. Rather than the relative strength of the direct absorption spectroscopy, WMS strategy is used for the measurement of a harmonic signal, which is proportional to the gas concentration. Thus, it is independent of the background spectrum and improves noise rejection.
- **Strong extensibility.** On the basis of non-dispersive infrared (NDIR), TDLAS, as a universal technique, has been extended to the measurement of enormous gases in various applications.
- **Strong anti-interference ability.** Different from PAS, TDLAS makes full use of pure spectrum absorption of an individual gas, and the noise interference can be ignored. It shows obvious advantages in terms of sensitivity, high zero stability, selectivity, high speed of response, and so on, proving a promising reference in the field of dissolved methane measurement of power transformer oil.

There are still some issues to be addressed in the line.

- **Long-term stability.** Further research can be focused on the expansion of more fault gas components and long-term reliability of the online optical monitoring system.
- **Sensitivity of the measurement system.** As for the TDLAS itself, the sensitivity is high, but the measurement performance of the system can be improved through extending the optical path, modifying the demodulation method, and using high-sensitivity infrared detectors, etc.
- **Cost of the manufacturing.** The high precision tunable diode laser and specially designed gas cell are costly, restricting the promotion of the online TDLAS devices.

4.5 Laser Raman Spectroscopy Technique

4.5.1 Detection Principle of Raman Spectroscopy

Laser Raman spectroscopy (LRS) is based on the phenomenon of the Raman effect, which was first discovered and proposed by C. V. Raman [66]. The simple principle of Raman scattering can be seen in Figure 4.49, as mentioned in Chapter 2. Different from near infrared spectroscopy, the properties and structures of uncertain matter can be investigated by measuring the Raman scattering light directly produced by the matter due to laser irradiation. More precisely, Raman spectroscopy provides detailed information about molecular vibrations. Extremely high sensitivity has been demonstrated for Raman spectroscopy in various industrial and agricultural fields [67].

Polyatomic molecules may have various vibration modes, corresponding to their respective Raman frequency shift spectral lines with different intensities. Thus the Raman characteristic spectral lines of the fault characteristic gases must be selected before performing qualitative and quantitative analyses.

The dissolved gases H_2 and CO in transformer oil are diatomic molecules, while CO_2, CH_4, C_2H_2, C_2H_4, and C_2H_6 are polyatomic molecules, which are separated for Raman spectrum analysis [32, 68–70].

Especially, H_2 detection based on absorption effects (absorption spectroscopy/PAS) has been considered infeasible for a long time. With the development of spectral detection technology and lasers, the phenomenon of H_2 electric quadrupole transition has been discovered, which provides the possibility for hydrogen detection based on absorption spectroscopy. According to the characteristic absorption spectrum of H_2, a

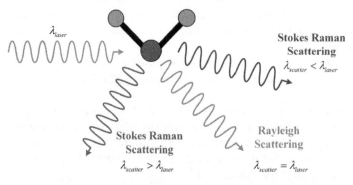

Figure 4.49 Schematic diagram of the Raman scattering phenomenon.

frequency-locking cavity enhanced absorption spectroscopy (FLCEAS) detection platform has been tried to realize the qualitative and quantitative detection and analysis [71].

4.5.2 Application of Laser Raman Spectroscopy

First of all, Raman spectroscopy of gases dissolved in power transformer oil should be clear. In order to solve the problem of mutual absorption interference during simultaneous detection of the characteristic gases dissolved in the transformer oil, it is necessary to determine a characteristic absorption spectrum of each gas. Taking hydrogen as an example, there are no absorption lines for the other dissolved gases (CO, CO_2, CH_4, C_2H_2, C_2H_4, C_2H_6, N_2, and O_2) at the specific position for H_2. The selection of the characteristic absorption spectrum of a certain gas should follow the following three main principles.

1. No overlap (more than twice the linewidth) of the absorption spectrum lines for different gases.
2. Independent characteristic absorption positions are suggested. The distance between the spectral line and other absorption spectral lines of the gas itself is greater than twice the line width to avoid mutual absorption interference due to the spectral line broadening effect.
3. High intensity at the absorption line of the characteristic spectrum is preferred in order to realize the trace of fault gases.

The Raman characteristic spectral lines of the seven fault characteristic gases in a transformer are determined by calculating. The chosen Raman characteristic spectral lines of the seven fault characteristic gases are shown in Table 4.7 [72].

It is recommended through the attributions of the Raman peaks of the transformer fault gases to eliminate the interferences of the external noise signal, which lays the foundation for measurement.

This method involves the analysis of Raman scattering, which is observed in samples when they are illuminated by a strong light source. There are two types of Raman spectroscopy techniques that are of interest for the evaluation of various substances. A typical Raman detection platform is illustrated in Figure 4.50.

Table 4.7 Spectral lines of the seven fault characteristic gases in transformer oil.

Gases	Characteristic frequency (cm^{-1})	Attribution of Raman peaks
H_2	4160	H—H bond stretching vibration
CH_4	2919	C—H bond symmetric stretching vibration
C_2H_2	1973	C—H bond and C—C bond symmetric stretching vibration
C_2H_4	1342	C—H bond scissoring vibration and C—C bond symmetric stretching vibration
C_2H_6	2954	C—H bond antisymmetric stretching vibration
CO	2144	C—O bond stretching vibration
CO_2	1387	C—O bond symmetric stretching vibration coupling with scissoring vibration to form Fermi peaks

Source: Data from Weigen et al. [72].

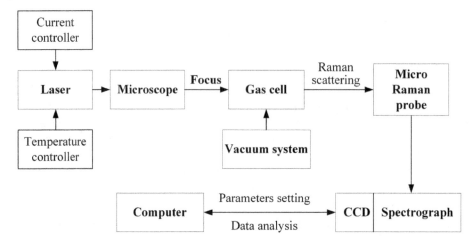

Figure 4.50 Raman detection platform for dissolved fault gases in transformer oil.

The simultaneous detection of seven kinds of dissolved gases, H_2, CO, CO_2, CH_4, C_2H_2, C_2H_4, and C_2H_6, has been achieved by using Raman spectra [72]. Moreover, the optimum average times of Raman detection increases the detection limits by a factor of more than 10 with the help of Allan variance analysis. Thus, the qualitative and quantitative analysis of LRS for dissolved gas in transformer oil is proved to be feasible in practice.

The rate of concentration of gases decomposed from transformer oil is fairly low, nearly close to the µL/L (ppm) level. On the other hand, Raman scattering cross-section of gases is quite small, as its intensity is generally only 10^{-6} of the incident light [72], which limits the application and development of Raman spectroscopy in the field of trace detection. Thus, it is necessary to enhance the sensitivity.

Surface-enhanced Raman Scattering (SERS) can effectively solve the problem of low sensitivity of Raman spectroscopy in trace analysis. At present, SERS is able to obtain the Raman fingerprint of the detecting materials with a 10^6 or higher enhancement factor in principle, especially with the advancement of nano-preparation technology and single-molecule level SERS detection. SERS has developed into a powerful and effective analysis tool, which is widely used in the field of heterogeneous catalytic reactions, the detection of explosives, the detection of environmental pollutants, etc.

SERS, which is an extension of Raman spectroscopy (RS), has been shown to enhance the Raman signals by several orders of magnitude by choosing suitable surfaces. It is foreseeable that surface-enhanced Raman spectroscopy has great application potential in detecting dissolved trace features in transformer oil. The key to SERS application is the preparation of a reliable SERS substrate with a high enhancement factor and good spatial uniformity. However, due to the low concentration of dissolved gases in the oil and a weak Raman signal, localized surface plasma resonance in the oil phase and preparation of surface-enhanced substrates are expected to achieve its successful application.

As a typical application, the detecting platform for dissolved gases based on silver nano-bulk surface-enhanced Raman spectroscopy is developing [73], as shown in Figure 4.51. In this process, inelastic light scattering by molecules is greatly enhanced when the molecules are adsorbed on to corrugated metal surfaces such as silver or gold

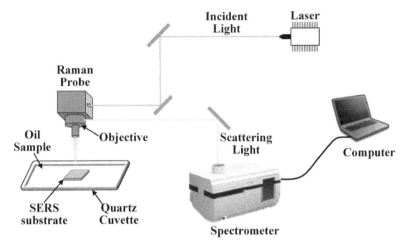

Figure 4.51 Schematic diagram of the Raman detecting optical path. Source: Modified from Gu et al. [73].

nanoparticles (NPs). Within decades, SERS has become a source of exciting scientific phenomena, as well as one of the most sensitive analytical techniques currently available.

Apart from non-confocal techniques like surface enhanced Raman spectroscopy, tip enhanced Raman spectroscopy, resonance Raman spectroscopy, etc., the confocal Raman technique delivers the best results in terms of spatial resolution and background suppression. Combined with a specifically designed high sensitivity gas sample cell, the confocal Raman technique might provide another way for online monitoring of dissolved gas in transformer oil.

4.5.3 Merits and Drawbacks

Compared with traditional on-line monitoring technologies, LRS has some distinguished advantages.

- **Strong applicability.** Raman spectroscopy is a general and widely used quantitative and qualitative technique to provide various substances with signature fingerprint spectra for the molecule under test, whereby dissolved gases in transformer oil can be traced, as well as other featured materials like furfurals.
- **Single-frequency laser.** With a single wavelength laser, dissolved gases in the oil can be excited and then detected by the Raman spectrum at the same time, without carrier gas consumption and separation of mixed gases.
- **Non-contact detection and small samples.** Raman detection is a non-contact and non-destructive measurement and does not consume samples. It is very useful for detecting substances in small samples. The detection speed is fast and continuous measurement can be realized.

However, there are some concerns to be addressed to meet the requirements for online detection of DGA in a power transformer.

- **Sensitivity improvement.** Since a Raman scattering cross-section of gas is quite small, enhancing the detection sensitivity for gas is the key application of LRS in a dissolved gas-in-oil analysis. Further improvements of the detection sensitivity are needed to achieve sensitivity limits suitable for application in power transformers in service.

- **Noise reduction.** In order to increase the gas detection limit, a smoothing de-noising technique like Allan variance can be used to improve the SNR of the measurement system apart from enhancement of the molecular Raman effect.
- **Onsite installation.** Precise components such as Raman spectrometers, silver-plated glass tubes, and Raman probes might not be recommended for use in transformer sites, so the optical topology and integrated design need to be further simplified. The composition of transformer oil is complex and additional efforts are still needed before it can be routinely used in commercial products.

4.6 Fiber Bragg Grating (FBG) Technique

4.6.1 Detection Principle of FBG

As mentioned in Chapters 2 and 3, fiber Bragg grating (FBG) is widely used in optical components as direct sensing elements for strain and temperature [74]. Also, it acts as a wavelength-specific reflector since the specific periodic variation in the refractive index of the fiber core, as depicted in Figure 4.52. The reflected Bragg wavelength λ_B, can be used as the demodulation signal instead of optical intensity.

It is interesting to engage FBG as a sensing component for the dissolved gas in transformer oil because of its novel design. The key is to generate a strain or temperature change from the measurand. An absorbent coating that expands in the presence of the target substance is necessary for a specific FBG gas sensor [75–77].

It has been noted that only hydrogen is achieved with the FBG approach using the available sensing membrane. It is therefore still not a general technique for all the feature gases dissolved in oil.

4.6.2 Application of the FBG Technique

4.6.2.1 Standard FBG Sensor

Experience shows that dissolved hydrogen is produced in most oil thermal and electrical faults inside power transformers. Hence it is meaningful to monitor the degradation

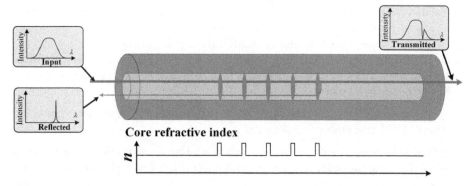

Figure 4.52 A fiber Bragg grating structure with a refractive index profile and spectral response.

process and health status of transformers by detecting dissolved hydrogen. Instead of the electronic sensor, the optical techniques provide different but excellent approaches to detect hydrogen by embedding the hydrogen sensors in the oil without inducing any insulation risks. The optical fiber hydrogen sensors can be divided into three main groups: interferometric-based, intensity-based, and fiber grating-based fiber-optic sensors. Taking advantage of its independence of light intensity, grating-based hydrogen sensors have been widely investigated in recent years [78–86]. As a suitable hydrogen sensor for online monitoring of power transformers, requirements consist of a limitation of detection (LOD) below 100 µL/L and a dynamic range of at least 2000 µL/L should be met. In addition, the sensor needs to detect in an oil temperature range from 50 to 80 °C.

Stress induced by volume expansion of Pd absorbing hydrogen results in a shift in the Bragg wavelength. In the presence of ambient hydrogen, Pd film absorbs H, and subsequently the stress induced by the volume expansion results in a swell of FBG; then a shift in the Bragg wavelength can be detected by the interrogator. In this way, hydrogen concentration can be calculated by noting the wavelength shifts. However, pure Pd film easily suffers from fatal fracture caused by its α–β phase transition and hysteretic effect. Alloying Pd with other metals such as Ag, structural stability of the hydrogen-sensitive film can be improved. The thick Pd film and Pd/Ag composite film are used to absorb hydrogen and the function of pure Pd film sputtered at the outermost layer is to stop oxidation of silver alloy and enhance sensitivity of hydrogen. The structure of the FBG-based hydrogen sensor is shown in Figure 4.53 [87]. Several layers have been made around the fiber cladding (125 µm): a polyimide layer (~1 µm), a titanium (Ti) layer (20 nm), a palladium/silver (Pd/Ag) composite layer (~400 nm), and a Pd layer (~160 nm).

In the design, Ti and polyimide layers act as adhesive coatings to ensure connection between fiber and the Pd alloy film.

The coating fabrication is also cautious to avoid uneven coating, so self-rotation of the sampling tray at a specific speed and a floating arrangement are necessary, as depicted in Figure 4.54. Magnetron dual-sputtering is preferred during preparation of the Pd/Ag composite film. The fabrication procedure of a typical FBG-based hydrogen sensor includes processing of FBG, a polyimide coating, and magnetron sputtering. The recommended parameters can be seen in Table 4.8.

To explore the sensitive performance, a comparative test, an FBG hydrogen sensor with a Pd/Ag composite film (Pd/Ag 400 nm, Pd 160 nm) and one with pure Pd film (560 nm), in transformer oil is carried out at the temperature of 60 °C. As the dissolved hydrogen in transformer oil increases, the wavelength of the FBG shifts more, in accordance with the sensing principle. The results of two FBG hydrogen sensor wavelength shifts are shown in Figure 4.55, where the red curve refers to the Pd/Ag composite film sensor and the blue curve refers to the pure Pd film sensor. Compared to the pure Pd

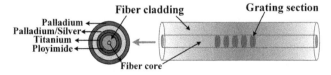

Figure 4.53 Different layers of the FBG-based hydrogen sensor. Source: Reprinted with permission from Ma et al. [87]. © 2015 American Institute of Physics.

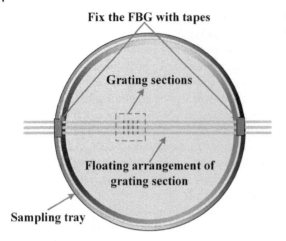

Fix the FBG with tapes

Grating sections

Floating arrangement of grating section

Sampling tray

Figure 4.54 FBG arrangement on the sampling tray.

Table 4.8 Detailed parameters of magnetron-sputtering coating.

Target material	Pd	Ag	Ti
Start power (W)	50	40	30
Sputtering power (W)	60	20	30
Start time (min)	10	10	5
Sputtering time (min)	23	15	5
Vacuum (Pa)	5×10^{-4}		
Working gas	Ar		
Gas flowrate (sccm)	20		
Working pressure (Pa)	1.0		
Deposition rate (nm/min)	20	7	4
Film thickness (nm)	~560		20

Source: Reprinted with permission from Ma et al. [87]. © 2015 American Institute of Physics.

film, the Pd/Ag film is more sensitive by about 25% for detecting dissolved hydrogen in oil under the same conditions.

To evaluate the mechanical performance, the morphology prepared by the field emission scanning electron microscope (FE-SEM) is compared after repetitive tests, as shown in Figure 4.56. Protuberances and cracks can be observed after tens of absorption cycles in hydrogen; in contrast, Pd/Ag had a better performance on inhibition of hydrogen embrittlement. Therefore, the FBG hydrogen sensor's reversibility and reliability can be improved by using Pd/Ag and pure Pd composite films.

Since the solubility of hydrogen dissolved in the oil is not dependent on oil temperature, it also makes a difference to the performance of the sensors. In order to provide a continuous monitoring of the health status of the transformer, the sensor needs to be examined at different temperature specifications. Thus, the oil temperature was set to range from 20 to 80 °C by adjusting the PID (proportion integration differentiation) temperature controller, which simulates the transformer working temperature and operational conditions, respectively. To observe the different temperature influences on the

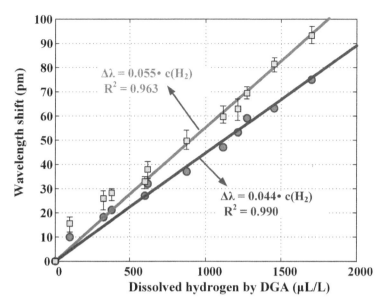

Figure 4.55 Sensitivity test at different hydrogen concentrations in oil. Source: Reprinted with permission from Ma et al. [87]. © 2015 American Institute of Physics.

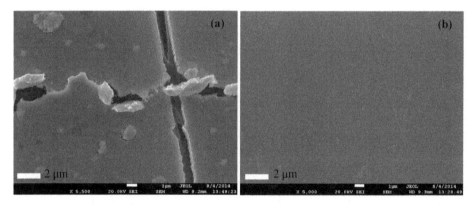

Figure 4.56 Comparison of SEM morphology between (a) a pure Pd film and (b) a Pd/Ag film. Source: Reprinted with permission from Ma et al. [87].

performance of the Pd/Ag sensor, the wavelength shifts of FBG sensors in oil samples with certain dissolved hydrogen at different temperatures are shown in Figure 4.57 in order, where the dotted lines stand for the steady state in different situations.

Then the wavelength shift and response time under increasing oil temperatures are slightly decreased, which are best described by an exponential function of the relationship, as shown in Figure 4.58, where T is oil temperature, tr means the response time, and $\Delta\lambda$ is the wavelength shift.

The response time, under the same temperature range, seems to be more vulnerable to temperature change than sensitivity. Although both sensitivity and response time are susceptible to temperature, the problem can be solved by stabilization of the sensor

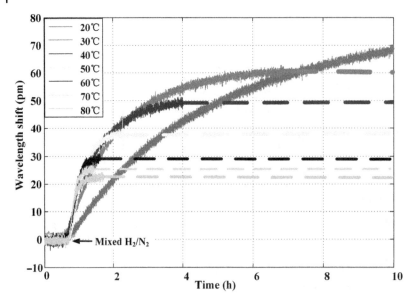

Figure 4.57 FBG hydrogen sensor with a Pd/Ag composite film at 20–80 °C. Source: Reprinted with permission from Ma et al. [87]. © 2015 American Institute of Physics.

Table 4.9 Specifications of various gas sensors for the detection of hydrogen.

Sensing coatings	Structures	Temperature	Limit of detection	Detection range	References
Pd/Ag	Thin film	20–120 °C	100 ppm		[88]
Pd, Au/Pd	Thin film	Room temperature	–	2–100%	[89]
Pd	Foil	90 °C	–	20–3200 ppm	[90]
Pd/Ag/Polyimide	Composite film	20–80 °C	<20 ppm	0–1800 ppm	[87]
Pd/Ag/Ti	Composite film	Room temperature	17 ppm	0–900 ppm	[91]
Pd/Ag/Ti	Composite film	Room temperature	2.1 ppm	0–700 ppm	[92]
$WO_3–Pd_2Pt–Pt$	Thin film	25°C	20 ppm	20–23 900 ppm	[93]
Pd, $Pd_{58}Cr_{42}$	Thin film	Room temperature	–	0–643 ppm	[75]
Pd	Thin film	Room temperature	–	0–719.7 µl/l	[94]
PTFE–Pd capped Mg–Ti	Thin film	21–80 °C	5 ppm	5–1500 ppm	[95]

Source: Modified from Sun et al. [1].

temperature and utilization of the fitted curves. Thus, the proposed hydrogen sensor can be expanded to monitor health status under an on-site situation.

Apart from the mentioned scheme, some other coating and structures have been tested and prepared; the performances are listed in Table 4.9 [1].

4.6.2.2 Etched FBG Sensor

To obtain satisfactory sensitivity, the Pd alloy sensing membrane has been demonstrated to be an effective technique, and etched cladding is seen to be an alternative approach

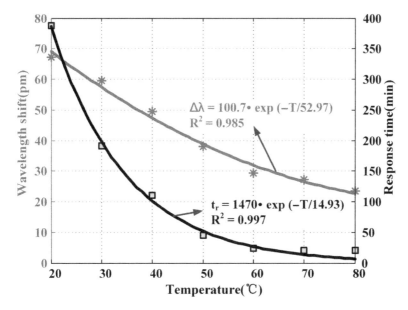

Figure 4.58 The fitting curves of oil temperature vs. response time. Source: Reprinted with permission from Ma et al. [87]. © 2015 American Institute of Physics.

Figure 4.59 Simplified model of the FBG-based hydrogen sensor. Source: Reprinted with permission from Jiang et al. [91]. © 2015 American Institute of Physics.

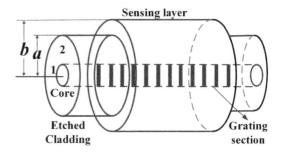

to improve the sensitivity [82, 96–99]. Specific emphasis is placed in this section on the effect of etching cladding, including the mathematical analysis, repeatability, and sensitivity test in mixed gases and transformer oil; a simplified model is shown in Figure 4.59.

In theory, when the sensor is exposed to hydrogen, the grating period of the sensor increases due to expansion of the Pd coating. The shift in Bragg wavelength can be calculated as [91]

$$\Delta \lambda_B = 0.026 \sqrt{ps} \left[\frac{(b^2 - a^2)Y_{pd}}{a^2 Y_F + (b^2 - a^2)Y_{pd}} \right] \times 0.78 \lambda_B \tag{4.12}$$

where $\Delta \lambda_B$ is the shift in Bragg wavelength with strain; λ_B is the Bragg wavelength of the FBG without any applied strain; p is the hydrogen partial pressure (torr); Y_{pd}, Y_F are the Young's modulus of Pd and fiber respectively (Pa), $Y_{pd} = 1.7 \times 10^{11} Pa$, $Y_F = 7 \times 10^{10} Pa$; and a, b are the radius value of cladding and sensing layer from the center of the fiber core (µm).

Then the wavelength shift ratio of etched fiber to standard fiber can be expressed as

$$k = \frac{b^2_{etch} - a^2_{etch}}{a^2_{etch}Y_F + (b^2_{etch} - a^2_{etch})Y_{pd}} \bigg/ \frac{b^2_{std} - a^2_{std}}{a^2_{std}Y_F + (b^2_{std} - a^2_{std})Y_{pd}} \tag{4.13}$$

where a_{std}, b_{std} are the radius values of the standard fiber cladding (μm) and a_{etch}, b_{etch} are the residual radius values of the etched fiber cladding (μm).

From the rough calculation and relationship curve, the fact that the smaller diameter of the fiber cladding brings out greater wavelength shifts under the same thickness of Pd membrane is confirmed.

To etch the diameter, 49% of HF acid solution is used to reduce the thickness of the fiber cladding in the process of preparing for etched FBG (EFBG) hydrogen sensors. Three diameters of EFBG sensors are fabricated with the cladding radii of 62.5, 54.5, and 46.5 μm respectively. A typical structure of an un-etched FBG sensor is illustrated in Figure 4.60.

Under the same hydrogen concentration in the mixed N_2/H_2 gases, the measured response time and wavelength shift are shown in Figure 4.61.

With regard to the identical hydrogen concentration, the EFBG sensor with a cladding radius of 46.5 μm is more sensitive than the un-etched sensor. To confirm the detection

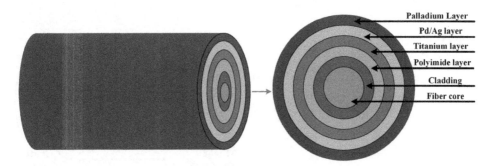

Figure 4.60 Structure of the EFBG-based hydrogen sensor.

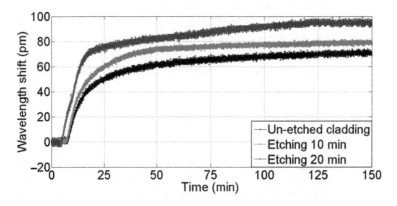

Figure 4.61 Hydrogen detection with three different cladding diameters. Source: Reprinted with permission from Jiang et al. [91]. © 2015 American Institute of Physics.

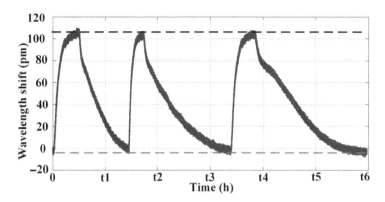

Figure 4.62 Wavelength shift of an FBG-based hydrogen sensor in the repeatability test. Source: Reprinted with permission from Jiang et al. [91]. © 2015 American Institute of Physics.

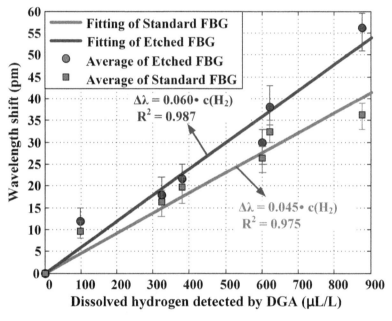

Figure 4.63 Contrasting sensitivity test of standard FBG vs. etched FBG in transformer oil. Source: Reproduced with permission from Jiang et al. [91]. © 2015 American Institute of Physics.

performance, two more tests in gas and transformer oil were carried out to verify the repeatability and sensitivity, as shown in Figures 4.62 and 4.63.

Compared with a standard FBG hydrogen sensor, the chemically etched FBG sensor has its own characteristics. On the one hand, the etched FBG sensor is more sensitive to the dissolved hydrogen in transformer oil. This is expected as the mathematical analysis is also similar to the test in mixed gases. On the other hand, the fluctuation of detection is worse than the standard FBG sensor. Due to etched cladding, which is very sensitive to the ambient variation and is mechanically fragile, a properly designed package may be an effective tool to solve the problem and keep the advantage of sensitivity to dissolved hydrogen in transformer oil.

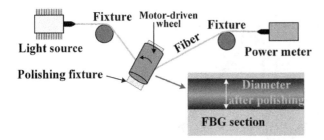

Figure 4.64 Fabrication arrangement of the side-polished FBG.

4.6.2.3 Side-Polished FBG Sensor

To achieve higher sensitivity, the side-polished FBG (SP-FBG) technique can be applied to further improve the performances [83, 100–106].

To fabricate the side-polished structure, the optical cladding of FBG is recommended to be mechanically polished by a motor-driven wheel, which straightens the FBG and adjusts the distance between fixtures at the ends of the fiber. During the side-polishing process, speed and stress of the wheel should be regulated. Meanwhile, the FBG was connected to a light source and an optic power meter, which helped to enable on-line monitoring of the polishing process, as shown in Figure 4.64. It should be noted that a special shelter arranged on the side-polished surface contributes to the control of sputtering just on the FBG section.

Similarly, sensing membrane can be magnetron sputtered on the side-polished surface. The structure of the FBG-based hydrogen sensor is illustrated in Figure 4.65. Here, r represents the radius of the standard fiber and the value is 62.5 μm and h is the residual thickness and the value is 20 μm. Alloying Pd with Ag (Pd/Ag, Ag 25%) helps to prevent potential embrittlement, offering resistance to poisoning and providing a longer operational lifetime.

To estimate the proper thickness of the sensing film, the simplified and ideal physical model of an SP-FBG hydrogen sensor is built with rectangular coordinates, as shown in

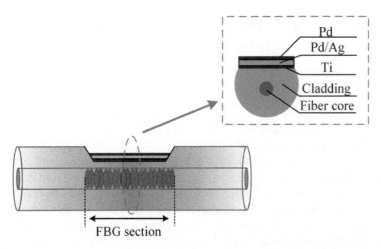

Figure 4.65 Structure of the SP-FBG hydrogen sensor in a lateral and cross-sectional view.

Figure 4.66 Simplified physical model of the SP-FBG hydrogen sensor.

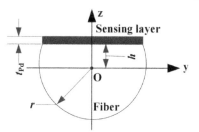

Figure 4.66. **O** is the center of the optical fiber. In this model, Pd/Ag composite film is simplified to one sensing layer of pure Pd (the Ti layer is ignored in this physical model due to its relatively thin thickness). Suppose that t_{Pd}, the thickness of the sensing layer, is less than 1 μm.

As the sensor relies on the strain effect induced on the fiber through this swelling of the sensitive coating, the key point is to obtain the relationship between the strain of Bragg grating and specific hydrogen concentration.

As Pd absorbs hydrogen, it expands, producing stress (strain) in the bulk materials. Strain in a free-standing bulk Pd is equal to the ratio of the fractional change in the lattice constant of the Pd. In the α-phase, the relationship between the strain of Pd and hydrogen partial pressure is given by [107]

$$\delta = 0.026\frac{\sqrt{p}}{K} \tag{4.14}$$

where p is the hydrogen partial pressure (torr), K is the Sievert coefficient ($K = 350\,\text{torr}^{-1/2}$), and p (torr) equals 7.6 times $c(H_2)$ in % [108].

The next step is to calculate the strain of bending deformation in the SP-FBG core due to a one-sided expansion of the Pd coating.

The sectional inertia moment of the SP-FBG sensor head consists of two parts: I_{Fy}, the inertia moment of the fiber and I_{Sy}, the inertia moment of the sensing layer. In our model, they are respectively expressed as

$$I_{Sy} = \int_A z^2\,dA = \int_h^{h+t} z^2 \cdot 2\sqrt{r^2 - h^2}\,dz = \frac{2}{3}\sqrt{r^2 - h^2}(t_{Pd}^3 + 3h \cdot t_{Pd}^2 + 3h^2 \cdot t_{Pd}) \tag{4.15}$$

$$I_{Fy} = \int_A z^2\,dA = \int_{-r}^h z^2 \cdot 2\sqrt{r^2 - z^2}\,dz = \frac{r^4}{4} \cdot \arcsin\frac{h}{r} + \frac{\pi r^4}{8} + \frac{r^4}{16} \cdot \sin(\arcsin\frac{h}{r}) \tag{4.16}$$

where r is the radius of the standard fiber and the value is 62.5 μm, h is the residual thickness and here the value is 20 μm, and t_{Pd} is the thickness of the sensing layer (μm).

Owing to the thickness of the sensing layer, which is much less than the radius of the fiber, it will not make a difference to leave out the inertia moment of the sensing layer. Thus, the inertia moment of the SP-FBG sensor is given as

$$I_y = \sum_{i=1}^n I_{yi} = I_{Sy} + I_{Fy} \approx I_{Fy} \tag{4.17}$$

The Young's modulus can be calculated as follows:

$$Y_{SP-FBG} = \frac{Y_{Pd} \cdot \sqrt{r^2 - h^2} \cdot t_{Pd} + Y_F A_F}{A_{Pd} + A_F} \approx \frac{Y_{Pd} \cdot \sqrt{r^2 - h^2} \cdot t_{Pd} + Y_F A_F}{A_F} \tag{4.18}$$

where Y_{Pd}, Y_F are Young's modulii of Pd and fiber, respectively (Pa), and A_{Pd}, A_F are the areas of Pd and fiber, respectively (μm^2).

Thus, the classic formula for determining the bending strain in the fiber under simple bending is [109]

$$\varepsilon = \frac{M \cdot z}{Y_{SP-FBG} \cdot I_y} \approx \frac{\delta \cdot A_F \cdot Y_{Pd} \cdot 2\sqrt{r^2 - h^2} \cdot t_{Pd} \cdot (h + \frac{1}{2}t_{Pd})}{I_{Fy} \cdot (Y_{Pd} \cdot \sqrt{r^2 - h^2} \cdot t_{Pd} + Y_F \cdot A_F)}$$

$$= 0.072 \cdot \frac{t_{Pd} \cdot (2h + t_{Pd})}{Y_{Pd} \cdot \sqrt{r^2 - h^2} \cdot t_{Pd} + Y_F \cdot A_F} \cdot \frac{Y_{Pd} \cdot A_F \cdot \sqrt{r^2 - h^2}}{I_{Fy}} \cdot \frac{\sqrt{c(H_2)}}{K} \quad (4.19)$$

where M is the bending moment (N·μm) and z is the perpendicular distance to the neutral axis.

According to Eq. (4.19), the strain of fiber Bragg increases with the thickness of the sensing layer. Thus, a thicker layer of Pd is beneficial to more wavelength shifts at the same hydrogen concentration. To calculate the relationship between the relative bending strain and thickness of the Pd coating, the following parameters are used: $r = 62.5\,\mu m$, $h = 20\,\mu m$, $Y_{Pd} = 1.7 \times 10^{11}$ Pa, $Y_F = 7 \times 10^{10}$ Pa, $A_F = 8591.1\,\mu m^2$, and $I_{Fy} = 6\,315\,010.7\,\mu m^4$. From the calculation, the strain induced by bending is expressed as fractions of the maximum concentrations of different thicknesses in Figure 4.67 (the maximum strain is expressed as 1 PU), where a linear curve is obtained with regard to the ideal physical model. Nevertheless, permeability of hydrogen would get worse when the Pd coating is thicker. Moreover, hydrogen embrittlement would occur on the thick film. In addition, the side-polished fiber is fragile with so much metal sputtered on the surface.

It is really important to evaluate sensing performances with different thicknesses of Pd/Ag films, but the proportion of Pd and Ag and the thickness distribution of Pd and Pd/Ag composite films will make a difference as well. Tanking all of the factors into consideration, around 560 nm of Pd/Ag composite film is recommended.

The developed SP-FBG sensor is then tested directly in transformer oil. A schematic diagram of the dissolved hydrogen experimental setup is shown in Figure 4.68. Low

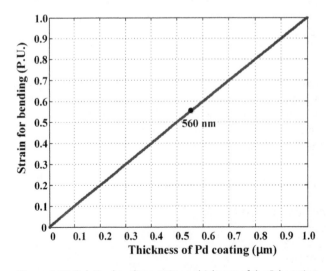

Figure 4.67 Relative bending strain vs. thickness of the Pd coating.

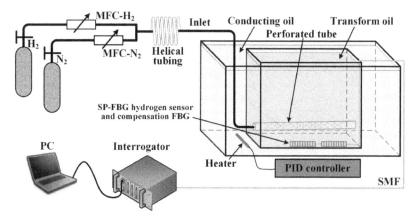

Figure 4.68 Experimental setup of dissolved hydrogen detection in transformer oil.

concentrations of hydrogen gas are obtained by adjusting the MFC (mass flow controller) flowrates and helical tubing is used to make the H_2/N_2 fully mixed. Mixed gases are used in a multi-step to change and limit the dissolved hydrogen concentration in order to protect the sensor from failure. A perforated tube with millions of tiny pores on its surface make it ideal for dissolving hydrogen in the oil and can stop transformer oil from flowing out. A PID controlled device is engaged to keep the temperature of the conducting oil and transformer oil as stable as possible.

To investigate the sensitive and repeatable performance of a developed hydrogen sensor, three tests are carried out in turn with a scheduled time interval of a fixed number of days. The data and fitting curve between the wavelength shift and hydrogen concentration are illustrated in Figure 4.69, where the wavelength shifts have been compensated due to temperature variations.

Repeatability is also a significant parameter for a sensor, which is used to describe the ability of a sensor to provide the same result, under the same circumstances. Standard deviations (shown in green) of wavelength shifts are depicted in Figure 4.70.

The SP-FBG sensor is improved using a side-polished technique and one-side surface coating where the sensitivity of the SP-FBG dissolved hydrogen sensor is 0.477 pm/(µL/L). Resolution of the interrogator SM-130 is 1 pm, which reveals that the resolution of the SP-FBG hydrogen sensor can be as good as 2.1 µL/L, which is much better than the standard FBG sensor with the same coating setting. The following reasons can explain the phenomenon. Firstly, because of the asymmetric structure of the side-polished FBG with a one-side surface coating, the deformation is performed as a curvature instead of an axial stretch, and as the FBG is laterally polished it has an intrinsic sensitivity to curvature. Moreover, compared with the standard FBG, the sensitive film sputtered on the SP-FBG is much closer to the fiber core and gratings section, and hence volume expansion is easier to bring in higher strain and wavelength shifts.

4.6.3 Merits and Drawbacks

In summary, an FBG-based dissolved hydrogen sensor provides a direct and effective detection of the health status of power transformers with several distinguished features.

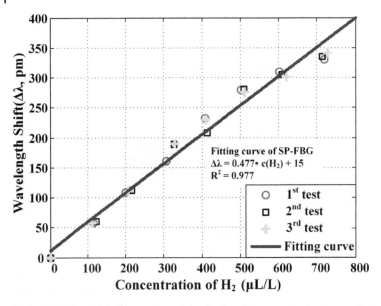

Figure 4.69 Sensitivity test results at a low hydrogen concentration in transformer oil. Source: Adapted with permission from Jiang et al. [92]. © 2015 Institute of Electrical and Electronics Engineers.

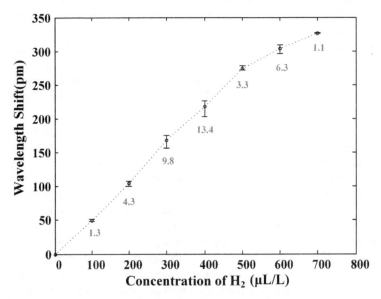

Figure 4.70 Error bars and standard deviations of wavelength shifts at low hydrogen concentrations. Source: Reprinted with permission from Jiang et al. [92]. © 2015 Institute of Electrical and Electronics Engineers.

- **Direct measurement.** Being different from the conventional DGA technique, the FBG scheme gets rid of the gas separation process from the oil. The FBG-based sensor can detect the dissolved hydrogen directly and even locate the defect or partial discharge quickly and easily.
- **Continuous and fast response.** The response time is an important characteristic for online monitoring in a power transformer. The FBG sensor can accomplish the

Table 4.10 Comparison of conventional gas chromatography and the FBG-based sensing method in oil.

Items	FBG-based technique	Conventional DGA
On-line monitoring	Sustainably	Periodically
Oil/gas separation	No need	Necessity
Carrier gases	No need	Necessity
Consumption of gas to-be-measured	Scarcely	Yes
Distributed measurement	Yes	No
Anti-electromagnetic interference	Excellent	Common

measurement within hours, skipping the oil–gas separation process. Fortunately, high temperature accelerates the response time to even less than 1 hour. Moreover, since there is no oxygen in transformer oil, the palladium on the sensors immersed in the oil is not easily oxidized and fails to absorb hydrogen.

- **Simple installation.** Oil–gas separation relies on complex mechanical components and carrier gases, so the direct FBG sensor is easy to install. The FBG method provides multiple sensing points shared by one single interrogation system, providing a relatively easy installation and an affordable price for the complete solution.

In addition, due to obvious advantages in wavelength division multiplex (WDM) and immunity to intensity variations of the light source, FBG-based hydrogen sensors are convenient to use to discover fault locations. A detailed comparison can be seen in Table 4.10.

However, there are still some problems to be solved.

- **Limited gases.** FBG-based sensors rely on the sensing material and membrane fabrication, which up to now have been limited to the utilization of dissolved hydrogen detection. More work is expected to be done on a breakthrough in dealing with related gas-sensitive materials.
- **Calibration of environmental conditions.** Temperature and oil flow have some impact on the sensitivity of the equipment, so accurate calibration needs to be determined for the industrial application in power transformers for different scenarios.
- **Long-term stability.** Although Pd alloy relieves the problem of hydrogen embrittlement, accumulated defects in the long term need more consideration and tests.

4.7 Discussion and Prediction

4.7.1 Comparison of Optical Fiber Techniques

Generally speaking, there is no perfect way to detect the dissolved gas in transformer oil. The concentration of some specific dissolved gas is a ratio of the percent concentration of the gas in the oil. In fact, the Ostwald coefficient, a measure of the solubility of a gas in a liquid, varies with temperature and pressure. It is not easy to obtain the exact measurement, but the data of DGA is vital to an in-service power transformer. Optical solutions may provide insightful solutions to the application of online DGA measurement.

As the optical technology has the obvious advantage of non-contact measurement, anti-electromagnetic interference, and free calibration, it is considered to have a bright prospect. In recent years, domestic and foreign scholars have carried out a variety of effective exploration and research. The technique of dissolving gas in transformer oil using an optical sensor mainly includes PAS, LRS, infrared absorption spectroscopy, and fiber grating sensing techniques. Compared with the traditional non-optical detection methods, optical techniques have several obvious merits in common.

(1) Non-contact measurement, as optical methods do not consume any gases separated from transformer oil; (2) the performance of the components is stable; (3) anti-electromagnetic interference; (4) there is no need to separate the gas column, through spectral analysis, to determine the gas composition and content, as detection speed can achieve continuous measurement.

Therefore, optical gas detection is an important direction to break through the bottleneck of traditional technology.

On the other hand, we should pay more attention to problems and challenges posed by optic-based methods. The challenges can be real motivations from a scientific and technical perspective. As to the above-mentioned optic-based DGA techniques discussed in this chapter, the brief advantages and disadvantages can be compared, as shown in Table 4.11, according to the different principles.

Table 4.11 Comparisons of various optic-based gas detection techniques for power transformer oil.

Physical process	Optical technique	Main challenges
Absorption (no radiation transition)	Photoacoustic spectroscopy	Relatively mature scheme, easily affected by noise and the micro-microphone quality and different gases of cross-interference
Scattering (non-elastic scattering)	Raman spectroscopy	Single-wavelength laser, weak intensity, cross-sensitive, fluorescent interference, low sensitivity
Transmission	Absorption spectroscopy	Cross-sensitive, susceptible to environmental parameters
	TDLAS	Highly sensitive, complex design and high cost
Interference	FTIR	Available for unknown substances in the laboratory, but difficult for on-line monitoring, susceptible to displacement drift (vibration), cross-sensitive
Reflection	FBG	Simple, easy to be installed, no need for gas/oil separation, only works for dissolved hydrogen

Existing optical techniques for the detection of dissolved gases in transformer oils still suffer from the following common disadvantages:

1. Cross-sensitive problem. In view of the symbol gases dissolved in transformer oil, there are variable degrees of cross-sensitive problems in the optic-based techniques. The wavelength resolution used in FTIR and the photoacoustic technique is roughly on nm-scale. Typically, the infrared absorption spectrum width of acetylene is about 22 pm and that of methane is about 35 pm, both of which are far less than the wavelength resolution used in the optical technique. Especially in the near-infrared section, the absorptive wavelengths of seven fault gases overlap a lot. Although the separation algorithm and multi-modal fitting can alleviate the cross-sensitive problem to a certain extent, it still does not solve the cross-sensitive problem fundamentally.

2. Insufficient sensitivity. The existing optical gas sensing techniques are widely used in the atmosphere, environment, and on special industrial occasions, with most of the applications demanding that the detection sensitivity is not very high. Although the measuring requirement on dissolved gases is at the ppm level, this means that simple transplantation of an optical sensing solution cannot meet the sensitivity standard for online DGA in transformer oil. On another aspect, some optic-based techniques have a very high sensitive performance in the laboratory and the effective sensitivity of the measurement system needs more consideration, particularly in practical application.

3. Dependence on the oil/gas separation process. The existing optical techniques mainly focus on the gas sensors themselves under circumstances of mixed gases that are completely dependent on the oil/gas separation. However, the traditional oil/gas procedure is the main factor restricting the detection cycle of the online monitoring. The dispersion of degassing is the main source of measurement error. In addition, the mechanical device for oil and gas separation makes the DGA equipment become very complicated.

4. Practical installation and interferences. The challenges from the practical application are tricky. For example, the noise in the field and the interference of the device itself affects high sensitivity detection for online monitoring. Also, the FTIR scheme is vulnerable to vibration and flutter issues.

4.7.2   Future Prospects of Optic-Based Diagnosis

Accurate measurement and good communication are the cornerstones of dissolved gases analysis relying on novel optical techniques. On the other hand, the detection results are just a kind of source data to be further analyzed. Some drawbacks of the DGA have been underscored by The Institute of Electrical and Electronics Engineers (IEEE) Std. C57.104, as follows: "Many techniques for the detection and the measurement of gases have been established. However, it must be recognized that analysis of these gases and interpretation of their significance is at this time not a science, but an art subject to variability." Further, "The result of various ASTM testing round robins indicates that the analytical procedures for gas analysis are difficult, have poor precision, and can be wildly inaccurate, especially between laboratories." Finally, "However, operators must be cautioned that, although the physical reasons for gas formation have a firm technical basis, interpretation of that data in terms of the specific cause or causes is not an exact science, but is the result of empirical evidence from which rules for interpretation have been derived."

Many years of empirical and theoretical study have gone into the analysis of transformer fault gases. Such, improved interpretation strategies are supposed to be carried out as well, to establish the effective relationship with the status of oil-immersed power transformers. Then, the condition-based maintenance will be more useful and the maintenance cost is sharply reduced and controlled.

DGA diagnosis techniques presented thus far use fault gas concentrations or ratios based on the practical experience of various experts, rather than on quantitative evidence. Now, with the availability of extensive DGA data, researchers are motivated to develop an alternative approach to DGA data interpretation alone. These different or alternate approaches include artificial intelligence (AI) techniques, fuzzy logic, and neural networks techniques.

Recent development of the AI model based on a combination of several techniques shows some future insight. AI approaches provide more accurate and reliable transformer diagnoses than a traditional algorithm alone. However, even though a majority of the AI approaches can diagnose faults with high accuracy, some of them fail to distinguish between thermal faults in oil and the same faults in cellulose, so engineering judgment is still required.

With regard to a power transformer to be tested or detected, it is necessary to match its actual condition with the exact features extracted from the case library, and use the corresponding case information to evaluate the health states of the transformer. If one is in the sub-health state, the historical information should be combined to carry out a short-term prediction to avoid severe casualty.

To realize a dynamic and early warning of the transformer, it is necessary to establish a model based on the characteristic fault gas data of the transformer and estimate the state of the device based on the established model parameters. Thereafter, a state change trend model can be established from the DGA data to predict the duration time. There is no need to calculate the time span of a healthy state when the transformer is first put into operation. However, it is necessary to further subdivide the operating state of the transformer into a healthy state, a sub-healthy state, and a faulty state. The sub-healthy state reflects the transitional state of the transformer from a healthy state to a faulty state, as shown in Figure 4.71.

It is useful to predict the duration of the sub-health state and provide dynamic warning of when condition-based maintenance can be achieved. As an instance, the Gaussian mixture model (GMM) can be established to find the characteristic gases of typical transformer faults and operating states. The main difficulty in learning Gaussian mixture models from unlabeled data is that it is usually unclear which points locate health, sub-health, or fault status. Since the time information is ignored in GMM, the incipient fault prediction is not available to merely rely on this model. Therefore, a time series algorithm can be taken into consideration for fault prediction and preventive maintenance. It has been proved that dynamic early warning and incipient fault prediction in sub-health status is useful for in-service power transformers.

At last, the dynamic fault prediction is able to provide a decision-making basis for practical condition-based operation and maintenance using the combination of novel optic-based techniques and advanced algorithms.

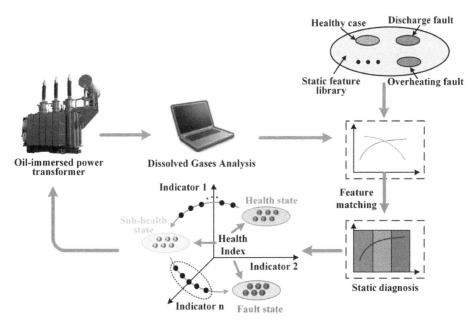

Figure 4.71 Illustration of health diagnosis and dynamic prediction of oil-immersed power transformers based on DGA. Source: Adapted with permission from Jiang et al. [110].

4.8 Conclusions

For the oil/paper insulation in a power transformer, events like arcing, corona discharge, sparking, and overheating bring in typical fault featured gases dissolved in oil. These gases can be detected in a transformer insulating oil using sensitive and reliable DGA techniques to determine the type of pending or occurring fault. Dissolved gases analysis for in-service transformers are accepted worldwide as one of the most effective methods to judge electrical or thermal faults inside oil-immersed power transformers.

It is easy and mature to measure the dissolved gas in the laboratory or offline, but the key to achieve transformer condition maintenance is online and involves analyzing oil-dissolved gases of operating transformers. Conventional online DGA devices suffer from a complicated structure, poor stability, a high failure rate, aging, and a heavy maintenance workload, while optical techniques become hot topics and promise alternatives for online monitoring due to excellent immunity to electromagnetic, eliminated chromatographic columns, and unnecessary carrier gases or calibration gases, in which chromatographic columns are easily contaminated and undergo aging thus requiring frequent replacement.

The technology of dissolved gas detection in transformer oil mainly includes PAS, Raman spectroscopy, spectroscopy absorption, and fiber grating sensing technology. Among these, the PAS solution has been manufactured and industrialized, while Raman spectroscopy, spectral absorption, and fiber grating techniques are still in the research stage. The advantages and disadvantages for each optical measurement method have been discussed and compared in this chapter. Accordingly, existing problems and future research directions have been outlined. Critical factors in various techniques were analyzed and investigated, such as the light source, long path gas cell, and the topology.

Unique properties and flexibilities in the application of optical sensing strategy for DGA enabled to-be-measured gases to be detected repeatedly without consumption and also made maintenance-free operation expectable. On the basis of accurate and effective optic-based measurement of dissolved gases, smart optical DGA with intelligent algorithms is going to be a powerful tool to evaluate the health status information of insulation oil in power transformers.

References

1 Sun, C., Ohodnicki, P.R., and Stewart, E.M. (2017). Chemical sensing strategies for real-time monitoring of transformer oil: a review. *IEEE Sensors Journal* 17 (18): 5786–5806.

2 Bustamante, S., Manana, M., Arroyo, A. et al. (2019). Dissolved gas analysis equipment for online monitoring of transformer oil: a review. *Sensors* 19 (19): 4057.

3 Ward, S. (2003). Evaluating transformer condition using DGA oil analysis. In: *Conference on Electrical Insulation and Dielectric Phenomena, 2003. Annual Report*, 463–468. IEEE.

4 Wan, F., Du, L., Chen, W. et al. (2017). A novel method to directly analyze dissolved acetic acid in transformer oil without extraction using Raman spectroscopy. *Energies* 10 (7): 967.

5 Varan, M. and Yurtsever, U. (2017). Multi-DGAS: a pattern based educational framework design for power transformers faults interpretation and comparative performance analysis. *Computer Applications in Engineering Education* 26 (2): 215–227.

6 Moravej, Z. and Bagheri, S. (2015). Condition monitoring techniques of power transformers: a review. *Journal of Operation and Automation in Power Engineering* 3 (1): 71–82.

7 Morais, D.R. and Rolim, J.G. (2006). A hybrid tool for detection of incipient faults in transformers based on the dissolved gas analysis of insulating oil. *IEEE Transactions on Power Delivery* 21 (2): 673–680.

8 Haema, J. and Phadungthin, R. (2013). A prediction technique of power transformer condition assessment via DGA parameters. In: *Power and Energy Engineering Conference (APPEEC), 2013 IEEE PES Asia-Pacific*, 1–4. IEEE.

9 Chatterjee, A., Bhattacharjee, P., Roy, N.K., and Kumbhakar, P. (2013). Usage of nanotechnology based gas sensor for health assessment and maintenance of transformers by DGA method. *International Journal of Electrical Power & Energy Systems* 45 (1): 137–141.

10 Bakar, N.A., Abu-Siada, A., and Islam, S. (2014). A review of dissolved gas analysis measurement and interpretation techniques. *IEEE Electrical Insulation Magazine* 30 (3): 39–49.

11 AnXin, Z., Xiaojun, T., Junhua, L., and Zhonghua, Z. (2014). The on-site DGA detecting and analysis system based on the Fourier transform infrared instrument. In: *Proceedings of the 2014 IEEE International Instrumentation and Measurement Technology Conference (I2MTC)*, 1036–1040. IEEE.

12 Aizpurua, J.I., Catterson, V.M., Stephen, B.G. et al. (2018). Power transformer dissolved gas analysis through Bayesian networks and hypothesis testing. *IEEE Transactions on Dielectrics and Electrical Insulation* 25 (2): 494–506.

13 Abu-Siada, A. and Hmood, S. (2015). A new fuzzy logic approach to identify power transformer criticality using dissolved gas-in-oil analysis. *International Journal of Electrical Power & Energy Systems* 67: 401–408.

14 Yangliu, L. and Xuezeng, Z. (2011). Estimation of dissolved gas concentrations in transformer oil from membranes. *Electrical Insulation Magazine, IEEE* 27 (2): 30–33.

15 Wang, X.F., Wang, Z.D., Liu, Q. et al. (2015). Dissolved gas analysis of a gas to liquid hydrocarbon transformer oil under thermal faults. In: *19th International Symposium on High Voltage Engineering*, Pilsen, Czech Republic, 424–429.

16 Liu, C., Zhang, H., Xie, Z. et al. (2019). Combined forecasting method of dissolved gases concentration and its application in condition-based maintenance. *IEEE Transactions on Power Delivery* 34 (4): 1269–1279.

17 Arakelian, V.G. (2004). The long way to the automatic chromatographic analysis of gases dissolved in insulating oil. *IEEE Electrical Insulation Magazine* 20 (6): 8–25.

18 Duval, M. and Dukarm, J. (2005). Improving the reliability of transformer gas-in-oil diagnosis. *Electrical Insulation Magazine, IEEE* 21 (4): 21–27.

19 Han, Y., Ding, F., Hao, C. et al. (2012). The oil–gas separation characteristics of ceramic/Teflon AF2400 composite membrane. *Separation and Purification Technology* 88: 19–23.

20 DL/T 1432.2-2016, *Testing Specification for On-line Monitoring Device of Transformation Equipment – Part 2: On-line Monitoring Device of Gases Dissolved in Transformer Oil*, National Energy Administration, China, 2016.

21 IEEE *Guide for the Interpretation of Gases Generated in Oil-Immersed Transformers*, 2009.

22 GB/T 7252-2001, *Guide to the Analysis and the Diagnosis of Gases Dissolved in Transformer Oil*, General Administration of Quality Supervision, Inspection and Quarantine of the People's Republic of China, China, 2001.

23 IEEE Power and Energy Society, *IEEE Guide for the Interpretation of Gases Generated in Mineral Oil-Immersed Transformers*, IEEE Std C57.104™-2019, United States, 2019.

24 Qi, B., Zhang, P., Rong, Z., and Li, C. (2020). Differentiated warning rule of power transformer health status based on big data mining. *International Journal of Electrical Power & Energy Systems* 121: 106150.

25 Dai, J., Song, H., Sheng, G., and Jiang, X. (2017). Dissolved gas analysis of insulating oil for power transformer fault diagnosis with deep belief network. *IEEE Transactions on Dielectrics and Electrical Insulation* 24 (5): 2828–2835.

26 International Electrotechnical Commission, *Mineral Oil-Impregnated Electrical Equipment in Service-Guide to the Interpretation of Dissolved and Free Gases Analysis*. International Standard IEC, 60599, 1999.

27 Zellweger, C., Emmenegger, L., Firdaus, M. et al. (2016). Assessment of recent advances in measurement techniques for atmospheric carbon dioxide and methane observations. *Atmospheric Measurement Techniques Discussions* 9 (9): 4737–4757.

28 Kalathripi, H. and Karmakar, S. (2017). Analysis of transformer oil degradation due to thermal stress using optical spectroscopic techniques. *International Transactions on Electrical Energy Systems* 27 (9): e2346.

29 Li, B., Zheng, C., Liu, H. et al. (2015). Development and measurement of a near-infrared CH4 detection system using 1.654 µm wavelength-modulated diode laser and open reflective gas sensing probe. *Sensors and Actuators B: Chemical* 225: 188–198.

30 Yu, Y., Sanchez, N.P., Griffin, R.J., and Tittel, F.K. (2016). CW EC-QCL-based sensor for simultaneous detection of H2O, HDO, N2O and CH4 using multi-pass absorption spectroscopy. *Optics Express* 24 (10): 10391–10401.

31 Li, C., Dang, J., Li, J. et al. (2016). A methane gas sensor based on mid-infrared quantum cascaded laser and multipass gas cell. *Guang pu xue yu Guang pu fen xi = Guang pu* 36 (5): 1291–1295.

32 Kamieniak, J., Randviir, E.P., and Banks, C.E. (2015). The latest developments in the analytical sensing of methane. *TrAC Trends in Analytical Chemistry* 73: 146–157.

33 Ke, Y., Long, Z., Xiaosong, W. et al. (2015). Methane concentration detection system for cigarette smoke based on TDLAS technology. *Spectroscopy and Spectral Analysis* 35 (12): 5.

34 Leis, J. and Buttsworth, D. (2018). A robust method for tuning photoacoustic gas detectors. *IEEE Transactions on Industrial Electronics* 65 (5): 4338–4346.

35 Chen, K., Zhang, B., Guo, M. et al. (2020). All-optical Photoacoustic multi gas analyzer using digital fiber-optic acoustic detector. *IEEE Transactions on Instrumentation and Measurement* 69 (10): 8486–8493.

36 Mao, Z. and Wen, J. (2015). Detection of dissolved gas in oil-insulated electrical apparatus by photoacoustic spectroscopy. *Electrical Insulation Magazine, IEEE* 31 (4): 7–14.

37 Wang, J., Zhang, W., Liang, L., and Yu, Q. (2011). Tunable fiber laser based photoacoustic spectrometer for multi-gas analysis. *Sensors and Actuators B: Chemical* 160 (1): 1268–1272.

38 Ma, Y. (2018). Review of recent advances in QEPAS-based trace gas sensing. *Applied Sciences* 8 (10): 1822.

39 Yang, T., Chen, W., and Wang, P. (2020). A review of all-optical photoacoustic spectroscopy as a gas sensing method. *Applied Spectroscopy Reviews*: 1–28.

40 Dumitras, D.C., Petrus, M., Bratu, A.-M., and Popa, C. (2020). Applications of near infrared Photoacoustic spectroscopy for analysis of human respiration: a review. *Molecules* 25 (7): 1728.

41 N'cho, J.S. and Fofana, I. (2020). Review of fiber optic diagnostic techniques for power transformers. *Energies* 13 (7): 1789.

42 Skelly, D. (2012). Photo-acoustic spectroscopy for dissolved gas analysis: Benefits and experience. In: *2012 IEEE International Conference on Condition Monitoring and Diagnosis*, 29–43. IEEE.

43 Zhang, Q., Chang, J., Wang, F. et al. (2018). Improvement in QEPAS system utilizing a second harmonic based wavelength calibration technique. *Optics Communications* 415: 25–30.

44 Dong, L., Li, C., Sanchez, N.P. et al. (2016). Compact CH4 sensor system based on a continuous-wave, low power consumption, room temperature interband cascade laser. *Applied Physics Letters* 108 (1): 011106.

45 Ding, J., Li, X., Cao, J. et al. (2014). New sensor for gases dissolved in transformer oil based on solid oxide fuel cell. *Sensors and Actuators B: Chemical* 202: 232–239.

46 Gordon, I.E. et al. (2017). The HITRAN2016 molecular spectroscopic database. *Journal of Quantitative Spectroscopy and Radiative Transfer* 203: 3–69.

47 Jiang, J., Ma, G., Song, H. et al. (2016). Tracing methane dissolved in transformer oil by tunable diode laser absorption Spectrum. *IEEE Transactions on Dielectrics and Electrical Insulation* 23 (6): 8.

48 Zha, S., Ma, H., Zha, C. et al. (2020). Trace gas detection based on photoacoustic spectroscopy in 3-D printed gas cell. *Journal of Near Infrared Spectroscopy* 28 (4): 236–242.

49 Sur, R., Sun, K., Jeffries, J.B. et al. (2015). Scanned-wavelength-modulation-spectroscopy sensor for CO, CO_2, CH_4 and H_2O in a high-pressure engineering-scale transport-reactor coal gasifier. *Fuel* 150: 102–111.

50 Zhu, C., Chang, J., Wang, P. et al. (2015). Continuously wavelength-tunable light source with constant-power output for elimination of residual amplitude modulation. *Sensors Journal, IEEE* 15 (1): 316–321.

51 S. Zhang, J. Wang, D. Dong, et al., "Mapping of methane spatial distribution around biogas plant with an open-path tunable diode absorption spectroscopy scanning system," *Optical Engineering*, vol. 52, no. 2, p. 026203, 2013.

52 Song, F., Zheng, C., Yan, W. et al. (2017). Interband cascade laser based mid-infrared methane sensor system using a novel electrical-domain self-adaptive direct laser absorption spectroscopy (SA-DLAS). *Optics Express* 25 (25): 31876–31888.

53 Duval, M. (2003). New techniques for dissolved gas-in-oil analysis. *Electrical Insulation Magazine, IEEE* 19 (2): 6–15.

54 Jiang, J., Wang, Z., Ma, G. et al. (2019). Direct detection of acetylene dissolved in transformer oil using spectral absorption. *Optik* 176: 214–220.

55 Tang, X., Wang, W., Zhang, X. et al. (2018). On-line analysis of oil-dissolved gas in power transformers using fourier transform infrared spectrometry. *Energies* 11 (11): 3192.

56 Tang, X. et al. (2018). Identification and treatment approach for spectral baseline distortion in processing of gas analysis online by Fourier transform infrared spectroscopy. *Spectroscopy Letters* 51 (3): 134–138.

57 Liu Lixian, H.H., Andreas, M., and Xiaopeng, S. (2020). Multiple dissolved gas analysis in transformer oil based Fourier transform infrared photoacoustic spectroscopy. *Spectroscopy and Sepctral Analysis* 40 (3): –4.

58 Zhang, Q., Chang, J., Wang, Q. et al. (2018). Acousto-optic Q-switched fiber laser-based intra-cavity Photoacoustic spectroscopy for trace gas detection. *Sensors* 18 (1): 42.

59 Yang, L., Jia-nan, W., Mei-mei, C. et al. (2016). The trace methane sensor based on TDLAS-WMS. *Spectroscopy and Spectral Analysis* 36 (01): 4.

60 Li, C., Dong, L., Zheng, C., and Tittel, F.K. (2016). Compact TDLAS based optical sensor for ppb-level ethane detection by use of a 3.34 μm room-temperature CW interband cascade laser. *Sensors and Actuators B: Chemical* 232: 188–194.

61 Frish, M.B., Laderer, M.C., Smith, C.J. et al. (2016). Cost-effective manufacturing of compact TDLAS sensors for hazardous area applications. In: *SPIE LASE*, 97300P–97300P-9. International Society for Optics and Photonics.

62 Jiang, J., Zhao, M., Ma, G. et al. (2018). TDLAS-based detection of dissolved methane in power transformer oil and field application. *IEEE Sensors Journal* 18 (6): 2318–2325.

63 Cai, T., Gao, G., and Wang, M. (2016). Simultaneous detection of atmospheric CH_4 and CO using a single tunable multi-mode diode laser at 2.33 µm. *Optics Express* 24 (2): 859–873.

64 Guojie, T., Dong, F., Wang, Y. et al. (2015). Analysis of random noise and long-term drift for tunable diode laser absorption spectroscopy system at atmospheric pressure. *Sensors Journal, IEEE* 15 (6): 3535–3542.

65 Jiang, J., Wang, Z., Han, X. et al. (2019). Multi-gas detection in power transformer oil based on tunable diode laser absorption spectrum. *IEEE Transactions on Dielectrics and Electrical Insulation* 26 (1): 153–161.

66 Lizhi, Z., Weigen, C., Fu, W., and Jing, S. (2013). Laser Raman spectroscopy applied in detecting dissolved gas in transformer oil. In: *2013 Annual Report Conference on Electrical Insulation and Dielectric Phenomena*, 1145–1148. IEEE.

67 Lee, H., Cho, B.-K., Kim, M.S. et al. (2013). Prediction of crude protein and oil content of soybeans using Raman spectroscopy. *Sensors and Actuators B: Chemical* 185: 694–700.

68 Jin, W., Ho, H.L., Cao, Y.C. et al. (2013). Gas detection with micro- and nano-engineered optical fibers. *Optical Fiber Technology* 19 (6): 741–759.

69 Hodgkinson, J. and Tatam, R.P. (2013). Optical gas sensing: a review. *Measurement Science and Technology* 24 (1): 012004.

70 Chong, X., Kim, K., Li, E. et al. (2016). Near-infrared absorption gas sensing with metal–organic framework on optical fibers. *Sensors and Actuators B: Chemical* 232: 43–51.

71 Fu, W., Weigen, C., Pinyi, W. et al. (2017). Detection study of transformer fault characteristic gases based on frequency-locking absorption spectroscopy technology. *Proceedings of the CSEE* 37 (18): 8.

72 Weigen, C., Fu, W., Zhaoliang, G. et al. (2016). The research for Raman analysis of dissolved gases in transformer oil and optimization of quantitative detection. *Transactions Of China Electrotechnical Society* 31 (2): 8.

73 Gu, Z., Chen, W., Yun, Y. et al. (2018). Silver nano-bulks surface-enhanced Raman spectroscopy used as rapid in-situ method for detection of furfural concentration in transformer oil. *IEEE Transactions on Dielectrics and Electrical Insulation* 25 (2): 457–463.

74 Rao, Y.J. (1999). Recent progress in applications of in-fiber Bragg grating sensors. *Optics and Lasers in Engineering* 31 (4): 297–324.

75 Samsudin, M.R., Shee, Y.G., Adikan, F.R.M. et al. (2016). Fiber Bragg gratings hydrogen sensor for monitoring the degradation of transformer oil. *IEEE Sensors Journal* 16 (9): 2993–2999.

76 Knapton, A. (1977). Palladium alloys for hydrogen diffusion membranes. *Platinum Metals Review* 21 (2): 44–50.

77 M. Butler, R. Sanchez, and G. Dulleck, "Fiber optic hydrogen sensor." Technical Report, SAND-96-1133, Sandia National Labs, United States, 1996. https://www.osti.gov/biblio/251330.

78 Zhang, Y.-n., Peng, H., Qian, X. et al. (2017). Recent advancements in optical fiber hydrogen sensors. *Sensors and Actuators B: Chemical* 244: 393–416.

79 Sutapun, B., Tabib-Azar, M., and Kazemi, A. (1999). Pd-coated elastooptic fiber optic Bragg grating sensors for multiplexed hydrogen sensing. *Sensors and Actuators B: Chemical* 60 (1): 27–34.

80 Silva, S.F., Coelho, L., Frazão, O. et al. (2012). A review of palladium-based fiber-optic sensors for molecular hydrogen detection. *Sensors Journal, IEEE* 12 (1): 93–102.

81 Fisser, M., Badcock, R., Teal, P. et al. (2017). Palladium based hydrogen sensors using fiber Bragg gratings. *Journal of Lightwave Technology* 36 (4): 850–856.

82 Dai, J., Yang, M., Yu, X. et al. (2012). Greatly etched fiber Bragg grating hydrogen sensor with Pd/Ni composite film as sensing material. *Sensors and Actuators B: Chemical* 174: 253–257.

83 Xiaowei, D. and Ruifeng, Z. (2010). Detection of liquid-level variation using a side-polished fiber Bragg grating. *Optics & Laser Technology* 42 (1): 214–218.

84 Silva, S. et al. (2013). H_2 sensing based on a Pd-coated tapered-FBG fabricated by UV femtosecond laser technique. *Photonics Technology Letters, IEEE* 25 (4): 401–403.

85 M. Samsudin, Y. Shee, F. Mahamd Adikan, et al., "Fiber Bragg Gratings (FBG) Hydrogen Sensor for Transformer Oil Degradation Monitoring," *Sensors Journal, IEEE*, vol. 16, no. 9, pp. 1, 2016.

86 Dai, J., Zhu, L., Wan, G. et al. (2017). Optical fiber grating hydrogen sensors: a review. *Sensors* 17 (3): 577.

87 Ma, G.M., Jiang, J., Li, C.R. et al. (2015). Pd/Ag coated fiber Bragg grating sensor for hydrogen monitoring in power transformers. *Review of Scientific Instruments* 86 (4): 045003.

88 M. A. Butler, R. Sanchez, and G. R. Dulleck, "Fiber optic hydrogen sensor,"; Sandia National Labs., Albuquerque, NM (United States)SAND-96-1133; Other: ON: DE96012035; TRN: AHC29614%%58 United States 10.2172/251330 Other: ON: DE96012035; TRN: AHC29614%%58 OSTI as DE96012035 SNL English, 1996, Available at: https://www.osti.gov/servlets/purl/251330.

89 Ohodnicki, P.R., Baltrus, J.P., and Brown, T.D. (2015). Pd/SiO$_2$ and AuPd/SiO$_2$ nanocomposite based optical fiber sensors for H_2 sensing applications. *Sensors and Actuators B: Chemical* 214: 159–168.

90 M. Fisser, R. A. Badcock, P. D. Teal, et al., "Development of hydrogen sensors based on fiber Bragg grating with a palladium foil for online dissolved gas analysis in transformers," in *Proceedings of Society of Photo-Optical Instrumentation Engineers (SPIE), Optical Measurement Systems for Industrial Inspection X*, 2017, vol. 10329, p. 103292P: International Society for Optics and Photonics. https://doi.org/10.1117/12.2267091.

91 Jiang, J., Ma, G., Song, H. et al. (2015). Note: Dissolved hydrogen detection in power transformer oil based on chemically etched fiber Bragg grating. *Review of Scientific Instruments* 86 (10): 106103.

92 Jiang, J., Ma, G., Li, C. et al. (2015). Highly sensitive dissolved hydrogen sensor based on side-polished fiber Bragg grating. *Photonics Technology Letters, IEEE* 27 (13): 1453–1456.

93 Dai, J., Peng, W., Wang, G. et al. (2017). Improved performance of fiber optic hydrogen sensor based on WO$_3$-Pd$_2$Pt-Pt composite film and self-referenced demodulation method. *Sensors and Actuators B: Chemical* 249: 210–216.

94 Luo, Y.-T., Wang, H.-B., Ma, G.-M. et al. (2016). Research on high sensitive D-shaped FBG hydrogen sensors in power transformer oil. *Sensors* 16 (10): 1641.

95 Mak, T., Westerwaal, R.J., Slaman, M. et al. (2014). Optical fiber sensor for the continuous monitoring of hydrogen in oil. *Sensors and Actuators B: Chemical* 190: 982–989.

96 Dai, J., Yang, M., Yu, X., and Lu, H. (2013). Optical hydrogen sensor based on etched fiber Bragg grating sputtered with Pd/Ag composite film. *Optical Fiber Technology* 19 (1): 26–30.

97 Shivananju, B.N., Yamdagni, S., Fazuldeen, R. et al. (2014). Highly sensitive carbon nanotubes coated etched fiber Bragg grating sensor for humidity sensing. *Sensors Journal, IEEE* 14 (8): 2615–2619.

98 Jin-fei, D. and Hong-yan, F. (2007). Spectral characterization of long-period grating pair with etched fiber-cladding. *Journal of Zhejiang University (Engineering Science)* 41 (3): 537–540.

99 Coelho, L., de Almeida, J.M.M.M., Santos, J.L., and Viegas, D. (2015). Fiber optic hydrogen sensor based on an etched Bragg grating coated with palladium. *Applied Optics* 54 (35): 10342–10348.

100 Kwang-Taek, K. et al. (2007). Hydrogen sensor based on palladium coated side-polished single-mode fiber. *Sensors Journal, IEEE* 7 (12): 1767–1771.

101 Dai, J., Yang, M., Chen, Y. et al. (2011). Side-polished fiber Bragg grating hydrogen sensor with WO3-Pd composite film as sensing materials. *Optics Express* 19 (7): 6141–6148.

102 Tien, C.-L., Chen, H.-W., Liu, W.-F. et al. (2008). Hydrogen sensor based on side-polished fiber Bragg gratings coated with thin palladium film. *Thin Solid Films* 516 (16): 5360–5363.

103 Jang, H.S., Park, K.N., Kim, J.P. et al. (2009). Sensitive DNA biosensor based on a long-period grating formed on the side-polished fiber surface. *Optics Express* 17 (5): 3855–3860.

104 Bilro, L., Jordão Alberto, N., Sá, L.M. et al. (2011). Analytical analysis of side-polished plastic optical fiber as curvature and refractive index sensor. *Journal of Lightwave Technology* 29 (6): 864–870.

105 Wen, S.-C., Chang, C., Lin, C. et al. (2015). Light-induced switching of a chalcogenide-coated side-polished fiber device. *Optics Communications* 334: 110–114.

106 Dai, J., Yang, M., Chen, Y. et al. (2012). Hydrogen performance of side-polished fiber Bragg grating sputtered with Pd/Ag composite film. *Sensor Letters* 10 (7): 1434–1437.

107 Butler, M. and Ginley, D. (1988). Hydrogen sensing with palladium-coated optical fibers. *Journal of Applied Physics* 64 (7): 3706–3712.

108 Peng, T., Tang, Y., and Sirkis, J.S. (1999). Characterization of hydrogen sensors based on palladium-electroplated fiber Bragg gratings (FBG). In: *1999 Symposium on Smart Structures and Materials*, 42–53. International Society for Optics and Photonics.

109 Gere, J. and Timoshenko, S. (1997). *Mechanics of Materials*. In: . Boston, MA: PWS Publishing Company.

110 Jiang, J., Chen, R., Chen, M. et al. (2019). Dynamic fault prediction of power transformers based on hidden Markov model of dissolved gases analysis. *IEEE Transactions on Power Delivery* 34 (4): 1393–1400.

5

Partial Discharge Detection with Optical Methods

To ensure the supply and safety of power transmission and distribution, partial discharge (PD) measurements are widely used in the detection for possible insulation defects, preventing the in-service failures of power transformers. According to the by-product of PD activities, different diagnostic methods are developed to determine the defect types and their severity with the measured data, including ultra-high frequency (UHF) detection, acoustic emission (AE) detection, high frequency current transformer (HFCT) detection, and their combinations. Nevertheless, the conventional methods are not always effective in field applications due to the concerns of electromagnetic interference (EMI) and sensitivity. Instead, optical fibers have the merits of natural safety, immunity to EMI, and small volume, but it is of interest to investigate to improve the PD detection. For the power transformer, several investigations demonstrated the excellent characteristics of an optical fiber sensor immersed in oil. From consideration of the perfect insulation perspective, the optical sensors can be placed directly into the power transformers to get close to the potential PD source, enabling them to diagnose small defects with a higher sensitivity.

5.1 PD Activities in Power Transformers

A power transformer is a necessary piece of equipment in a power system and its reliability is related to whether the power grid can operate safely and steadily. The internal insulation failure of a transformer will have a significant impact on stable operation of it and will cause PD under the action of the electric field. The existence of PD will accelerate insulation deterioration in a transformer, even causing an electrochemical breakdown.

There are two main reasons for PD in the internal insulation of a transformer. First of all, in the long-term operation of power transformers, the insulating medium is prone to aging and moisture, and bubbles and electrode burrs appear inside the equipment, causing non-penetrating discharge in the high-voltage equipment. Secondly, operation of the transformer is affected by power system overvoltage, while the insulation defects will cause PD and accelerate insulation aging. Eventually, the transformer insulation will break down, so that the entire power system cannot operate normally or may even become paralyzed. PD with a short time has little effect on the normal operation of equipment, but if it is long and continuous, the insulation medium of the equipment will be damaged. The defective insulation medium will lose its protection function for high-voltage equipment. Furthermore, it is easy to result in power outages, fires and

Optical Sensing in Power Transformers, First Edition. Jun Jiang and Guoming Ma.
© 2021 John Wiley & Sons Ltd. Published 2021 by John Wiley & Sons Ltd.

other electrical accidents, and the life and property safety of the country and residents are threatened [1–4]. PD is generally the main reason for insulation problems in high voltage equipment, but it is also an important indicator for measuring the quality of the insulation medium [5, 6]. Therefore, it is necessary to monitor PD to reflect the working state of the transformer insulation.

There are two methods for monitoring PD of a transformer: offline monitoring and online monitoring. Offline monitoring is mainly to be carried out during planned power outage treatment for transformers, and to judge the equipment health states. If it cannot find the internal fault in time, the power failure maintenance consumes a lot of manpower and material resources. Online monitoring is mainly based on relevant equipment to achieve real-time monitoring of transformers. By analyzing the collected information in real time, the internal conditions of equipment can be obtained. As online monitoring can ensure safe and stable operation of the transformer and reduce the time of unnecessary power failure repairs, it has been widely employed and researched.

5.1.1 Online PD Detection Techniques

Generally, the PD detection in power transformers is based on various phenomena in the discharge process. The characteristics not only include movement of charge, the loss of electrical energy, and local overheating, but also involve electrical pulses, electromagnetic radiation, light, ultrasonic waves, etc. Thus PD detection methods can be accessed using the pulse current method, oil chromatography analysis method, UHF detection method, infrared detection method, and ultrasonic detection method, which are illustrated in Figure 5.1.

The pulse current method and oil chromatography analysis method belong to the offline monitoring methods. The insulation structure of the oil-immersed transformer is mainly composed of transformer oil and insulation paper. When PD occurs inside the transformer, insulation oil and insulation paper decompose and generate gas. Oil chromatography is used to determine whether there is a low energy discharge (PD fault) by analyzing the gas composition in insulating oil with the content of various gases. Diagnosing the fault using this method is difficult in the early stage of defect as it can

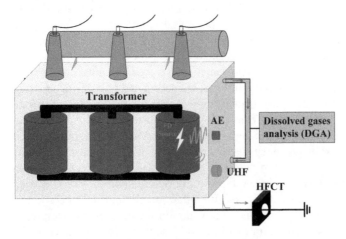

Figure 5.1 PD detection based on different approaches.

only detect the transformer on a regular basis. The pulse current method is based on an electric pulse signal and is the most conventional method used to detect PD. This method mainly captures the pulse currents, combining with data processing algorithms, to obtain PD information. The pulse current method has high accuracy and the apparent discharge can be measured, which is standardized by IEEE and IEC. However, it still has the deficiency of offline monitoring and concern about EMI.

The UHF method is a new method developed in recent years to overcome the shortcomings of interferences. If the PD occurs, the PD source within the transformer oil-paper structure radiates electromagnetic waves with a frequency domain up to GHz. At this time, the common environmental interference signals on site are generally lower than 400 MHz; thus the UHF method can effectively avoid the effects of ambient interference. Compared with the pulse current method, the UHF method can perform online monitoring with higher sensitivity and locate the PD source. At present, the development of UHF PD detection is mainly used for live detection. It does not affect the normal operation of equipment, but there are still some problems, such as the need for high resolution data acquisition (DAQ) and low detection efficiency.

The detected characteristic parameter of the ultrasonic detection method is PD-induced acoustic emission, which can suppress EMI, help online detection of PD, and locate the probable PD source. Thus, this solution has obvious advantages in PD detection.

5.1.2 PD Induced Acoustic Emission

It is apparent that electrical corona and high voltage spark make noise and produce acoustic emissions in a similar fashion, and acoustical methods of detecting and locating discharges have proved very valuable. The PD event is electrical in nature but is accompanied by vibration and excitation of ultrasonic signals that propagate through the insulation system. The discharge process is a relatively complicated physical process.

Due to PDs being generally generated by voids in oil or micro-gaps in solids, the generation of PD ultrasonic signals can be analyzed in a simplified bubble model, as shown in Figure 5.2, and there are two ways to produce a PD ultrasonic signal.

On the one hand, it is assumed that there is a bubble with a radius r and a mass M_m inside the insulation medium. When the bubble is in the electric field, it is maintained by the applied electric field force F_e and elastic force F_q. Ignoring the decay process, the PD is regarded as a single pulse signal. When it occurs, the electric field force disappears and the equilibrium state is destroyed. At low voltage, it still decays and oscillates, leading to generation of ultrasonic waves. On the other hand, the bubble in the oil medium is broken down at high voltage. The PD comes into being as a non-uniform spark channel

Figure 5.2 Simplified model of an insulation defect prior to PD.

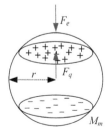

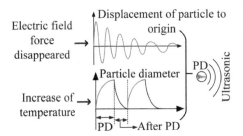

Figure 5.3 Mechanism of partial discharge ultrasonic signal generation.

with a small width. The discharge channel is strongly ionized by heating gas and reaches its maximum value rapidly. Then the energy is released as the electric field strength gradually becomes weak. Finally, this change in volume of the bubble generates sparse and dense waves outside as ultrasonic waves.

When ignoring the oscillation process of PD, the amplitude of ultrasonic wave generated by PD is proportional to PD intensity, as depicted in Figure 5.3. Based on this and the above analysis, the change of ultrasonic signal can reflect PD and monitor the insulation state of the equipment. Acoustical sensors, if correctly designed, are unaffected by electrical induced noise, and for this reason are very useful for verifying the existence of a discharge. Traditional PD ultrasonic signals are detected by acoustic emission (AE) probes with the functions of online monitoring, PD source location, pattern recognition, and quantitative analysis. These are now used routinely for detecting and locating discharges within power transformers and for finding radio interference sites caused by discharges associated with high-voltage power lines.

Ultrasonic signals of PD caused by an insulation defect were conventionally detected by acoustic emission (AE) sensors. While conventional acoustic emission monitoring is performed with the help of piezoelectric sensors mounted externally on the tank walls, they suffer from EMI issues and can only pick up a narrow band of frequencies due to their inherent resonances. In addition, the AE sensor with a metal material cannot be installed inside a power transformer for PD detection. At the same time, because of the compact internal structure of the transformer, an AE sensor can easily produce a large number of refraction shots and serious ultrasonic signal attenuations. The ultrasonic detection method that only relies on an AE sensor is difficult to ensure sensitivity and accuracy of PD detection. The signals in such sensors are prone to EMI while coupling from the sensor head to the measurement unit, thereby resulting in false alarms. Fortunately, the optical detection method is more advantageous when taking into account the requirement of EMI. Acoustic sensor-based optical fibers are EMI-free and can be installed inside the transformer tank walls in a minimally invasive manner. These fiber optic sensors may be broadly classified as fiber Bragg grating (FBG)-based detection, Fabry–Perot (FP)-based detection, and interference-based measurement, according to the optical components [7, 8]. In this chapter, various optical techniques are analyzed, and Fabry–Perot interferometer(FPI) and two-beam interference structures are considered separately to illustrate the application and relevant techniques, as shown in Figure 5.4.

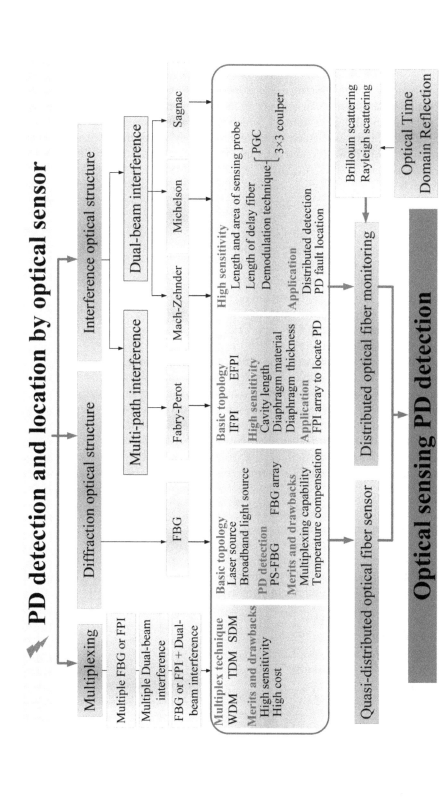

Figure 5.4 The optical PD detection using different techniques.

5.2 FBG-Based Detection

5.2.1 FBG PD Detection Principle

A FBG is essentially a sensor of temperature or strain, but by designing the proper inter-face, many other measurands can be made to impose a perturbation on the grating and produce a shift in the Bragg wavelength, which can then be used as a parameter trans-ducer. Therefore, FBGs are considered excellent sensor elements, suitable for measuring various engineering parameters such as temperature, strain, pressure, tilt, displacement, acceleration, load, as well as the presence of various industrial, biomedical, and chemical substances in both static and dynamic modes of operation. Since FBG is a cheap, stable, reliable, and widespread optical component, it attracts considerable interest to be used as a PD sensor [9–13].

As mentioned above, the weak ultrasonic effect can be the bridge to connect PD activ-ities with FBG sensors. The acoustic emissions generated due to PDs are typically so feeble that the resultant detected optical signal is in the sub-nW range. In addition, PD signals are wideband in nature, typically extending over tens of kHz up to 200 kHz. The principle of an FBG sensor is usually based on the relationship between the measurand and the Bragg wavelength. However, the frequency of the Bragg wavelength demodu-lation system is usually less than 5 kHz, and higher interrogation frequency devices are extremely expensive. Therefore, the transformer PD ultrasonic signal is not supposed to be obtained by directly demodulating the Bragg wavelength changes.

A significant challenge in using FBG-based sensors for sensing acoustic emissions is the effective conversion of wavelength encoded information to intensity variations. Hence, an appropriate technique to improve and guarantee the sensitivity of the detected signal needs to be implemented. At present, several techniques that have been demonstrated include using tunable narrow band lasers, edge filters, arrayed waveguide gratings, and interferometric methods [8, 14]. Wherein, two main topologies are recommended to demodulate the ultrasonic signal through an FBG element, according to the light sources.

- **Laser source.** A tunable laser source (TLS) or laser diode (LD)-based demodulating system of FBG can be employed for detection of acoustic emissions generated due to discharges, as shown in Figure 5.5. The FBG sensor connected to port 2 of the circulator is illuminated using a narrow band optical source, which sends light down the fiber. The optical signal reflected from the grating is captured by a high speed photodetector (port 3 of the circulator) and converts the light intensity to an electrical

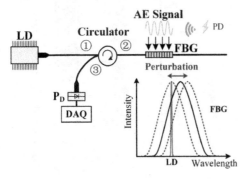

Figure 5.5 FBG detection systems using a narrow line-width laser diode.

signal. The key point is that the line width of the laser is much narrower than that of the FBG reflected wave. The light emitted from the TLS is biased and crosses to the reflection slope of the FBG. Then any change in Bragg wavelength caused by variations in strain/temperature is converted into an equivalent change in optical intensity since the reflectivity amplitude of the FBG is different at variable levels of perturbation [8, 13]. These optical intensity variations are detected using a photodetector, resulting in a corresponding change in voltage levels. A tunable laser-based FBG interrogation configuration converts the ultrasonic signal into an intensity of cross-points for a laser source and FBG.

It also should be noted that the upper limit on the incident acoustic frequency is set by the length of the grating. Specifically, the grating length should be less than half the acoustic wavelength in the fiber core.

- **Broadband light source.** A broadband light source, such as a light emitting diode (LED), is a cheap approach to achieve the demodulation and interrogation of an FBG-based ultrasonic signal [7]. A schematic diagram of a typical broadband light-based system is shown in Figure 5.6. Likewise, if an internal PD occurs in the electrical equipment, the ultrasonic wave causes the equipment to vibrate. The sensor will move and modulate the center wavelength of the reflected wave. In this manner, the operating point of the light moves and its amplitude also changes. Changes in the amplitude can then be converted to changes in electrical signals that can be detected by the photodetector, achieving detection of vibration signals [15, 16]. It is easy to understand that the bandwidth of the FBG is much narrower than that of the LED, but is similar to the case for laser versus FBG.

The methodology and detection of broadband light have been reported and demonstrated, which is the mainstream proposal. For demodulating the optical response signal, a tunable matching FBG is a good choice to be utilized [17]. The spectrum intensity tracks the wavelength shift to some extent, which is correlated to the amount of strain/temperature experienced by the FBG using suitable calibration [14].

One point that should be empathized is that the Bragg wavelength of FBG sensors is sensitive to both strain and temperature. In the scenario of PD detection, the temperature changes typically occur at a much slower rate (a few Hz) compared to acoustic emission (higher than 20 kHz). It seems that any temperature variation during the measurement of acoustic emission signals can be neglected. However, the temperature will have a significant impact on the work point of the FBG with the interactive function of light sources, and it is also necessary to take full consideration of the temperature effect.

Figure 5.6 FBG detection systems using a broadband light source (such as a light emitting diode, LED).

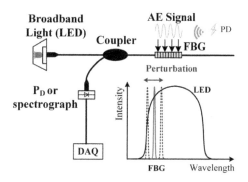

5.2.2 PS-FBG PD Detection

Up to now, the special type of FBGs whose reflection spectrum have a notch arising from a π(pi)-phase discontinuity in the center of the grating (PS-FBG, also called π-phase-shifted FBG) attracts intensive interest. This is because π-FBG may provide a solution to the sensitivity problem and demodulation strategy of the FBG with regard to highly sensitive ultrasonic detection. The grating exhibits a sharp resonance, whose centroid wavelength is pressure sensitive. By bringing a π-phase shift into a refractive index (RI) modulation of the FBG during the fabrication process, the spectral transmission has a narrow bandpass resonance appearing within the middle of the reflection lobe of the FBG, as shown in Figure 5.7. In contrast to standard fiber sensors, the high finesse of the resonance – which is the reason for the sensor's high sensitivity – is not associated with a long propagation length. Light localization around the phase shift reduces the effective size of the sensor below that of the grating and is scaled inversely to the resonance spectral width. For example, as to an effective sensor length of 270 μm, the effective bandwidth of 10 MHz was achieved [12].

Such an element allows a very narrow transmission band to be reached at the level of a few picometers (pm) and the sharp slope is very sensitive to weak perturbation induced by PD activities.

The cross-section point of a tunable laser wavelength and the reflected Bragg grating wavelength is expected to be located in a linear region. Then the response of the perturbation can be transferred into an accurate and linear measurement. In this sense, the cross-section point can be seen as a working state. Due to the PS-FBG, the waveform is fixed during the fabrication, and the wavelength of the laser source can be controlled and is tunable. The wavelength of the laser is prone to be set in the middle of the rising/falling edge of the reference grating. To have a better understanding, both linear regions are compared between normal FBG and PS-FBG with the same excitation, as illustrated in Figure 5.8. These are the operation points of the probe. Any drift away

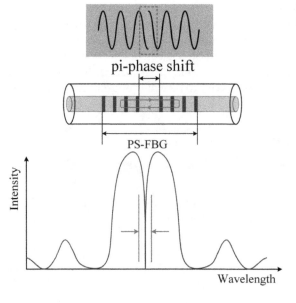

Figure 5.7 Illustration of a PS-FBG wavelength characteristic.

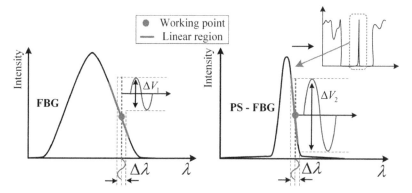

Figure 5.8 The linear region comparison between normal FBG vs. PS-FBG.

from this point is detected by the photodetector and therefore the acoustic emission induced change will be detected at the output.

It is obvious that a high change in intensity to the wavelength ratio gives high sensitivity and a narrow linewidth gives a high resolution with regard to PS-FBG. With the consideration of optical and electrical background noises in the measurement system, a high intensity of the laser also helps to give a high signal-to-noise ratio (SNR).

To verify the effectiveness of the PS-FBG PD detector, several cases have been examined in the laboratory [12, 13, 18, 19]. Typically, in order to detect the high frequency ultrasound signal, a PS-FBG ultrasonic detection system based on the tunable laser can be arranged both inside and outside of the transformer oil tank. As shown in Figure 5.9, the system consisted of PS-FBG probes, a tunable laser, optical coupler, photodetector, filter, and amplifier.

The narrow band laser source was emitted from the tunable laser with a power level of mW, a line-width level of 100 kHz or less and was launched into the FBG sensor via an optical circulator. Then the light reflected from the sensor was transmitted to the photodetector, converting optical signals into electrical signals. The ultrasonic signal with a certain frequency and a certain energy works on the FBG, which was compressed and stretched at the same frequency, causing the regular change of Bragg wavelength. In the meantime, the light intensity in the reflection spectrum of the FBG varies, leading to changes in the frequency and the peak-to-peak voltage amplitude of electrical signal collected by the DAQ module. In that sense, the ultrasonic signal is demodulated.

Specifically, the inside PS-FBG can be immersed into the oil directly since it has an excellent insulation advantage, and the outside PS-FBG needs to be attached to the shell so that it can sense the perturbation induced by the PD just like the conventional PZT (piezoelectric transducer) sensor.

Frequency response experiments indicate that the sensitivity of the PS-FBG is 8.46 dB higher than that of the conventional PZT sensor. Moreover, to test the sensitivity of the PS-FBG for on-line monitoring and on-site detection, a comparison experiment is performed. For the PS-FBG mounted to the outside surface of the oil tank, the sensitivity of the PS-FBG is 4.5 times higher than that of the PZT. After being immersed in the oil, the sensitivity of the PS-FBG is improved by 17.5 times [18].

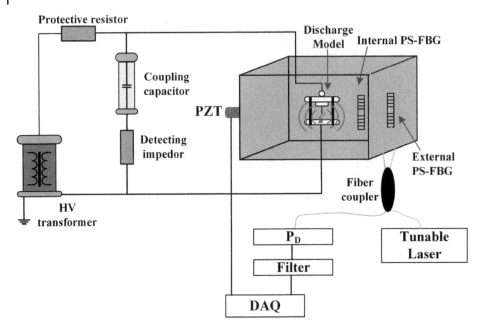

Figure 5.9 Schematic diagram of the PD detection platform based on a PS-FBG ultrasonic sensing system.

However, a PS-FBG owns a highly sensitive state in a quite narrow region and it is also vulnerable to ambient surroundings, mainly temperature. The sensitivity is prone to disappear once there is any drift away from this point or zone. Therefore, it is necessary to fixate the optimal state in the case of wavelength drifting. Conventionally, a similar FBG temperature sensor is mounted to be compensated for the extra wavelength shift due to temperature variance. The response is slow and the precision is low with regard to PS-FBG application. To solve the problem, the system can be optimized on the basis of direct wavelength scanning and compensation of temperature. After the direct wavelength scanning is implemented, the maximum scanning wavelength interval with an allowable error can be carried out on the basis of the cross-correlation algorithm calculation to improve the scanning efficiency, as displayed in Figure 5.10. Specifically, the wavelength offset is calculated and evaluated within the maximum wavelength interval, and the central wavelength of the tunable laser is scanned at precise intervals (e.g. 1 pm) to correct the actual optimal working wavelength of the light source. In this manner, acoustic emission sensing and development of an active tracking technique is established. Temperature fluctuation is compensated and the optimal working point of the tunable laser is guaranteed.

Robust classification of PDs in transformer insulation based on acoustic emissions detected using FBGs is carried out and proves that an FBG-based acoustic emission sensor system is capable of distinguishing the different types of PD events based on their frequency content. The ability to withstand a harsh environment coupled with having a smaller footprint and wide bandwidth make these sensors suitable for use in real transformers [8].

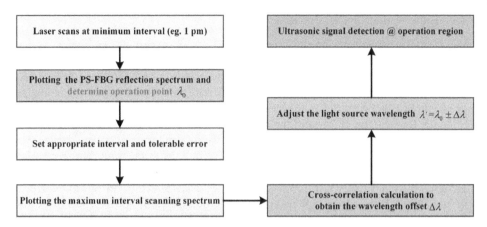

Figure 5.10 Strategy of temperature compensation-based cross-correlation algorithm.

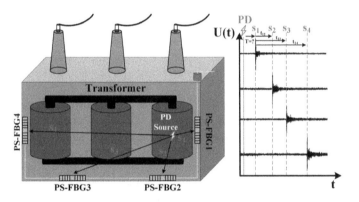

Figure 5.11 PD location with several PS-FBG sensors within an oil tank.

Furthermore, several PS-FBG sensors are arranged in a power transformer oil tank and a precise geometric PD position can be determined by the arrival times of the acoustic signals at each of the sensors, as shown in Figure 5.11. It is assumed that the PD pulse signal is detected instantaneously and the arrival time of the current pulse signal can be used as time zero for the PD event. As the time interval of the TLD wavelength reset is set to one second, a PD occurring once every 50 cycles can be detected (power frequency, 50 Hz). PDs generated by insulation defects repeat at each cycle. Therefore, distributed PD detection can be achieved via wavelength and time division multiplexing (TDM). The PD detection period for the entire power transformer equals the reset time interval multiplied by the PS-FBG numbers. Moreover, the acoustic system can work with an electric pulse detector to locate the insulation defect. The difference in the arrival times between the electric signal and an acoustic signal is the propagation time between the PD source and the sensor location. Using the time of flight (TOF) algorithm and known propagation times, the insulation defect can be located.

Since the higher ultrasonic response and closer installation location are within the oil tank, the PS-FBG methodology provides an ideal solution that substantially improves PD detection sensitivity in a power transformer.

5.2.3 High Resolution FBG PD Detection

Apart from the tunable laser technique, FBG PD detection can also be done with less limitation on the light source. Typically, a broadband laser is used as the laser source and, with a circulator, a feedback loop is designed for the photodetector to receive an optical signal [9, 10, 15, 16]. Especially, one FBG and another matched FBG is recommended, which is illustrated in Figure 5.12. The photodetector transforms an optical signal into an electrical signal and the data are recorded by a DAQ module or oscilloscope. In the design, FBG and the matched FBG have a similar reflection spectrum. The broadband light source enters the sensing FBG through the circulator and reflects back a narrow band signal light in Channel 1 (CH1). The signal light also passes through the matching FBG in CH2. Then the corresponding electrical signal of the photodetector is obtained. After the electrical signal is processed and filtered by software, accurate and reliable measurement information can be analyzed. The shaded part of the two FBGs is received optical power. When the reflection spectra of the sensing grating and the matching grating completely overlap, the light intensity reflected from the matching grating is the largest, which is equivalent to the maximum power. Assuming that the local ambient temperature of the two groups of gratings is unchanged and the axial strain of FBG changes, the center wavelength of FBG drifts a bit. When the temperature of the matching grating environment and the strain on the grating are known, the demodulation of the reflection wavelength at the center of the sensing grating can be achieved by monitoring the received intensity.

Inspired by the schematic, an experimental setup can be done with the following topology, as shown in Figure 5.13. The sensing measurement consists of two FBGs, which are attached to the outer and inner surfaces of the hollow mandrel shell, wherein the mandrel is mainly used as a PD acoustic amplifier. To cover the broad frequency bandwidth that is required for detection, the physical properties of the material, such as Young's modulus, Poisson's ratio, density, and the dimensions of the pipe should be taken into consideration during the fabrication. Teflon rods are installed inside the transformer model and inserted into the mandrel to keep them fixed.

The broadband laser passes through FBG1 and the transmits spectrum of FBG1 passes through FBG2. The reflected spectrum of FBG2 is selected by the fiber circulator. The acoustic signal pressure shifts the Bragg wavelength of the FBG. Due to the location of the FBG on the inner and outer parts of the mandrel shell, the wavelength shifts for FBG1 and FBG2 occurred in opposite directions. To get a proper operation region, the two FBGs must be selected such that they match the optical spectra zone. The transmitting

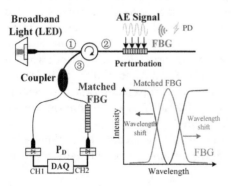

Figure 5.12 PD detection principle with two FBGs.

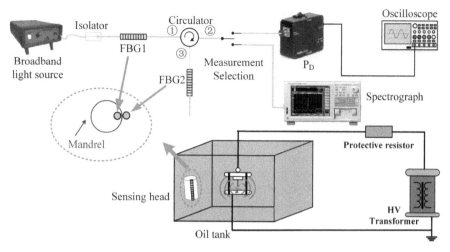

Figure 5.13 Experimental setup of PD detection using two FBGs. Source: Reprinted with permission from MDPI [10].

spectrum at the concave point and the reflected spectrum of the peak point occurred at the same wavelength when the sensor is in the initial state.

When the mandrel shell was concave, the outer FBG is stretched, the inner FBG is compressed, and vice versa when the mandrel shell is convex. Therefore, sensor sensitivity is doubled. The photodetector measures the area under the overlap of the spectra of the two FBGs. Because the spectra of the two FBGs shift in opposite directions, when the overlap of the spectra increases or decreases, the output voltage increases or decreases, respectively. A synchronization signal should be used to generate and detect PD signals. The synchronization signal was created by a function generator and used to trigger the PD generator and oscilloscope. The PD acoustic wave propagated through the oil and the sensor detected the acoustic signal. The PD signal is successfully detected and the PD frequencies are obtained in the frequency domain, which shows the validity of the results. A high-resolution sensor for PD-induced acoustic waves in transformer oil is verified with the combination of mandrel and FBG in terms of the frequency response, sensitivity, and direction and temperature dependencies [10].

5.2.4 Merits and Drawbacks

In nature, the fiber grating is stretched and shortened periodically at the frequency of the acoustic emissions induced by PD activities. The phenomenon provides an approach to sense the acoustic signal by various FBGs. It can be inferred that a single uniform fiber Bragg grating sensor is unsuitable for this kind of application due to the small Bragg shift to perturbation and relatively low commercial wavelength demodulation frequency. According to the choice of light source, mainly two types of FBG sensors are involved: a narrow laser source and a broadband source. Generally, a narrow laser source helps to achieve a high sensitivity on the perturbation since the laser has a narrow linewidth in a range of tens of kilohertz, but a proper and flexible design of sensing probes also make it possible to achieve PD detection on the basis of a broadband source at a relatively low cost. FBG-based PD measurements have many advantages.

- **Small size.** The diameter of a fiber is less than 1 mm and the length of a Bragg grating is usually less than 20 mm.
- **Low cost.** Basically, FBG is a common and universal component and the manufacturing cost is quite low, making it attractive to research and fabrication.
- **Multiplexing capability.** FBG sensor has a wonderful multiplexing capability, making it possible to achieve a distributed PD detection system in a real transformer tank. Some experiments are demonstrated. Moreover, it helps to improve the defect localization in a power transformer.
- **High sensitivity.** As mentioned in this chapter, there are several ways to improve the sensitivity of an FBG PD sensor. Using the PS-FBG, exhibiting a sharp resonance, the ultrasonic sensitivity is improved compared with the FBG. this performance makes such a design attractive for PD applications, in which compact, sensitive, and wideband acoustic measurement is available.
- **Immersing installation.** In addition, the general safety of the measurement system can be improved due to the insulating characteristics of optical fibers, while the integration with the emerging smart-grid technology can be easily achieved. Thus, FBGs are considered to be key devices and important alternatives when compared with conventional technologies such as strain gauges or solid-state sensing.

There are also some challenges on FBG-based sensors used for PD detection.

- **Interference on the acoustic signals.** In essence, the FBG-based PD sensor is transduced from the ultrasonic perturbation. The acoustic signals emitted due to the PD configurations are shaped by the immediate environment around which they are emitted. However, the present work is focused only on the methodology of classification. Although the sensitivity of acoustic measurement is not as good as the electrical/UHF measurement, the technique presented here is found to be quite robust. Even though aging of the insulation may create some changes in the discharge characteristics, the technique of classification will still hold good provided there are no drastic changes in the spectral characteristics. On the contrary, the ultrasonic frequency bandwidth of PD is concentrated to 20 kHz and above.
- **Temperature compensation.** There is flexibility in tuning the laser used for interrogating the grating prior to the measurement to compensate for any temperature-induced drift in Bragg wavelength. Temperature fluctuation is expected to be compensated and the optimal working point of the tunable laser should be guaranteed with an active tracking strategy.
- **Installation issues.** Future studies on this topic are required to establish the optimal sensor design and placement inside the transformer.

5.3 FP-Based PD Detection

5.3.1 FP-Based Principle

The Fabry–Perot (FP) optical structure is one of the typical optical interference structures. However, conventionally its additional FP cavity is different from other optical interference structures, which are entirely constructed within optical fibers. Therefore, the FP interference structure is discussed separately in this section.

The physicists Charles Fabry and Alfred Perot proposed the promising optical sensor, called the Fabry–Perot interferometer (FPI) sensor, which has advantages of precision, universality, and anti-EMI. Interesting and challenging fabrication methods have been used for the development of the optical sensors, such as vacuum evaporation, the micromachining technique, thin polymer film, and the ionic self-assembly monolayer (ISAM) technique, etc. According to a great deal of research on FP sensors, they have been used to realize online monitoring of structures, like detection of an aircraft jet engine for insulation defects and overheating faults [20], earthquake early warning, measurements in an oil well, navigation purposes such as optical fiber gyroscopes, sensing in biology, dynamics, vibration, and chemistry [21], as well as location and detection of a PD-induced acoustic wave [22, 23].

Traditionally, the PD ultrasonic signal and position can be obtained by an acoustic emission (AE) sensor (piezoelectric sensor) or its array, which are fixed in a transformer outside an enclosure. Due to the acoustic signal passing through the enclosure, it attenuates considerably and is vulnerable to the significant EMI in an actual high voltage power system, and the AE sensor suffers from a low SNR. To overcome the shortage of a conventional AE sensor, the optical fiber sensor based on FPI has been developed and illustrated. It can also achieve the PD location in a high voltage transformer [23]. The PD ultrasonic detection is one of the optical sensor's applications and has made wide and successful progress. The high sensitivity and small size of FPI sensors make them more suitable for PD detection and location. By adjusting the length of the sensing optical fiber, the size and sensitivity of an FPI sensor can be controlled and the phase resolution ratio of the high performance optical sensor reaches as high as 10^{-6} rad [24]. The Michelson and Mach-Zehnder interferometric structures were also designed successfully in applications of PD detection. However, a common problem between Michelson and Mach-Zehnder is that a long sensing length is required to guarantee sufficiency in detection [25]. Thus the FPI sensors array is preferred to find the position of the PD source and detect acoustic PD signal only with a short length of optic fiber.

The typical Fabry–Perot interference structure includes five parts: a laser light source, fiber coupler, sensor probe with an FP cavity, a photoelectric conversion module, and a signal output module; the connection of different devices is shown in Figure 5.14 [26]. The light with a single wavelength and narrow band emitted by a laser light source is propagated into the sensing probe through a fiber coupler and single-mode fiber. The first reflection of light that occurs at the end of the fiber core is in the FP cavity. A small amount of incident light returns along the original path of the fiber core and most of the optical signal undergoes a second reflection at the quartz film of the FP cavity. The second reflected light also turns back to the coupler through a single-mode fiber. To ensure the equality between lights of the second reflection and the first reflection, the tilt angle of the quartz film is adjusted. Then the two beams of optical signal with the

Figure 5.14 Illustration of atypical Fabry–Perot interference structure and the tropology. Source: Modified from Deng et al. [26].

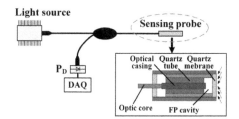

same amplitude, direction, and frequency interfere at the coupler. When there is a vibration signal from outside, the quartz film deforms, which results in a length change of the FP cavity and affects the phase of the second reflected light. Hence there is a phase difference between the lights of the first and second reflections. In summary, weak vibration, generated by ultrasonic signals of PD, is finally reflected through the laser light intensity detected by the photodetector.

The light intensity output by a photoelectric detector is shown as

$$I(\lambda) = \frac{R_1 + R_2 - 2\sqrt{R_1 R_2} \cos \varphi}{1 + R_1 R_2 - 2\sqrt{R_1 R_2} \cos \varphi} I_0(\lambda) \tag{5.1}$$

In Eq. (5.1), $I_0(\lambda)$ expresses the incident light intensity, λ is the wavelength of the light source, R_1 shows the reflectance of the optical fiber, R_2 shows the diaphragm reflectance, φ is the phase difference between two reflection lights, which can be represented as $\varphi = 4\pi l/\lambda$, and l is the length of the FP cavity. The length change of the FP cavity is caused by diaphragm deformation when there has been vibration of PD and results in a length change of the FP cavity. From Eq. (5.1), it can be seen that the output light intensity is only dependent on the FP cavity length. Thus the PD ultrasonic signal is transformed into light intensity vibration and is detected by a photodetector. That is to say, the impulse electric signal detected by the photodetector through the PD-induced acoustic emission and the light is just employed as the media to sense and transmits fault information [23].

5.3.2   Application of FP PD Detection

Due to the wide measurement range, high measurement accuracy, and low cost of the FPI sensor, it is widely used in research in the search for power apparatus. The development of FPI sensors has led to the emergence of multiple sensing structures and can be mainly divided into two types: the intrinsic Fabry–Perot interferometer (IFPI) and the extrinsic Fabry–Perot interferometer (EFPI).

The IFPI type is one of the earliest researched optical fiber FPI sensors and has attracted much attention since the 1990s. Figure 5.15 shows the typical structure of an IFPI. After removing the coating of the single-mode fiber, it is cut into three parts and the end face of the fiber is polished. The IFPI cavity is composed of an optical fiber being placed between the two fibers. Then a special film like TiO_2 is coated on both ends of this fiber and is fused with the other optical fiber to form an IFPI sensor. In addition, the IFPI sensor structure can be completed and processed by operation of laser writing. The two points on the sensitive fiber are exposed by focused ultraviolet rays. The refractive index will be changed permanently to form a virtual FP cavity when

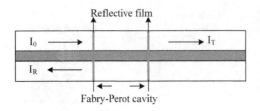

Figure 5.15 Schematic diagram of a typical IFPI sensor structure.

the position of the fiber core fluctuates. This method can effectively shorten the cavity length in a small range with the merit of low reflectivity.

Because the FP cavity in an IFPI is composed of sensing optical fiber, and light always transmits in an optical fiber, the optical wave loss caused by the coupling between the IFPI and a single mode fiber can be reduced. In addition, the loss of the interference process decreases simultaneously. After obtaining the interference spectrum of the IFPI reflected light, the numerical information of the FP cavity length and the core refractive index can be indirectly measured through the demodulation algorithm. However, the shortcomings of the IFPI sensors are that they are seriously disturbed by other physical quantities and the sensitivity of the single parameter is affected. That is, the single physical quantity is measured in an occasion with multiple physical information, and the multiple physical signals involved in IFPI and the single physical quantity are hard to measure. Therefore, it is not suitable to sense only the ultrasonic information using the IFPI sensor.

The IFPI sensor is obviously affected by temperature when it is used in the scenario of strain and other physical quantities. Thus, the IFPI sensor is not suitable to sense a single physical quantity. In response to the IFPI sensor shortcomings of a single sensitivity, EFPI offers another solution. A single-mode optical fiber with a smooth cut face is inserted into the collimated capillary. The sensor is very sensitive to a single physical quantity before interference from other physical quantities can be established. Because the interior of the FP cavity in an EFPI sensor is air or other media, not a sensitive optical fiber, its refractive index is not affected by lateral stress. Compared with the sensitive optical fiber, it can be applied, preferably, to measure physical quantities. At the same time, the length, material type, and sensing structure of the FP cavity inside the EFPI sensor significantly affect the measurement resolution and sensitivity. The EFPI sensor has attracted the attention of various institutions and has been widely used in the field of practical sensing. To satisfy the needs of different physical quantities and different dynamic measurement ranges, the structure of the EFPI sensor has been changed and optimized.

The most common structures in the FPI sensor include a thin diaphragm structure and a microcavity structure. The FPI sensor based on a diaphragm generally adopts corrosion or fusion to construct an air gap on the flat end surface of the optic fiber and an FPI cavity is formed by fixing a microsensitive layer on the surface. When external pressure acts on the diaphragm structure at the end of the fiber, the film is deformed, which causes length changes in the FP cavity. The stress sensing can be realized using the analyzing interference spectrum. The radius and thickness of the diaphragm of the EFPI sensor can be changed to improve range and sensitivity of the sensor, which is widely used in acoustic sensing and biomedical fields. The FPI sensor combined with a thin diaphragm structure has many advantages, such as high sensitivity, low manufacturing and maintenance costs, and a long detection distance. Due to the diaphragm, thickness is closely related to its stress sensitivity, so that reducing the diaphragm thickness and increasing its surface area can effectively improve the detection sensitivity of the FPI sensor. So far, there have been various methods of preparing FPI sensor probes with a thin diaphragm, such as focused ion beam (FIB) polishing, chemical etching technology, microelectromechanical system (MEMS) technology, and arc discharge technology [27–29].

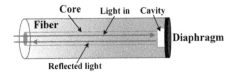

Figure 5.16 FP cavity processed by laser micromachining technology.

The laser micromachining process and nano-processed synthetic diaphragm technology are helpful in fabrication of a typical diaphragm for an FPI sensor. The main structure of the miniature optical fiber sensor processed by laser is all made of a quartz material and its typical structure diagram is shown in Figure 5.16. Laser processing of a microfiber sensor uses laser to etch the single-mode fiber, coating to form an FP cavity. A cylindrical mirror is adopted to collimate and shape the output light beam of the excimer laser into a spot. As a result, a square beam with uniform intensity distribution can be obtained by re-transmit. The optical fiber is fixed in a glass capillary tube and is etched on the surface by a square laser beam. The polycarbonate can be made as a sheet and is cured and broken at the end of the optical fiber to prepare the specific sensor.

Throughout the production process of the sensor, the stability laser intensity can be controlled by a computer for real-time feedback. The main difficulty is how to eliminate the surface reflection and the application of an ultraviolet curing adhesive.

The FPI sensor based on a microcavity structure generally measures a change in physical quantity indirectly according to the interference spectrum. Due to the more stable structure and higher measurement range, the EFPI sensor of a microcavity structure has a better performance in the field of acoustic and stress sensing in high temperature occasions. The microcavity FPI sensor that is made of optical fiber and other structures also has its unique advantages, such as a wide dynamic measurement range, simple preparation operation, and a stable sensor structure. The sensing structure based on an optical fiber microcavity is diverse, such as a built-in capillary optical fiber, a hydrofluoric acid etching multimode optical fiber, and splicing and etching of different types of optical fibers. Actually, the measurement sensitivity of the structure and its manufacturing technology has also been greatly improved.

The typical microcavity FPI sensor is formed by single-mode and multi-mode optical fibers, as shown in Figure 5.17. An all-fiber structure FP sensor fuses the hollow fiber and single-mode fiber obtained by first etching and then the single-mode fiber and multi-mode fiber are etched or stretched to ensure the same outside diameter. Alternatively, arc thermal fusion technology is used for fusion splicing the single-mode and multi-mode fibers, which can then be inserted into the hollow fiber. Then an FP cavity is formed between the light guide fiber (single-mode fiber) and the end of the multi-mode fiber. The white light interferometer and fine-tuning mechanism are used to monitor and adjust the length of the FP cavity. When the FP length reaches the design requirements, the inserted position of the single-mode and multi-mode fibers are fixed and welded to the hollow optical fiber.

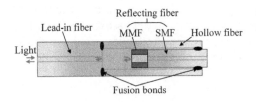

Figure 5.17 FP cavity composed of an all-fiber structure.

Since the main part of the all-fiber structure FPI sensor uses fiber material, it has good thermal stability. Meanwhile, it also has the advantages of not being affected by temperature, electromagnetic fields, corrosion, high temperature, etc. Compared with the fiber-optic diaphragm sensor, it is easier to manufacture and is able to have anti-friction and a wide detection range. However, due to processing limitations, the size of the all-fiber structure FP cavity sensor is slightly larger than the other types.

In addition, FPI sensors are very sensitive to thermal variations [30, 31] and are demonstrated to sense the change of thermal conditions and direction according to thermal sensitivity [32]. Transformer temperature is an important factor in determining its operational risk and service life. The rise of transformer temperature exceeding the limit accelerates insulation aging, damage to insulation, reduces service life, and can even cause power accidents. To transmit and detect the information of temperature, FPI sensors have been optimized in different ways; for example, the composite temperature sensor of an optical fiber can be obtained by combining the Michelson method and FPI sensors [33]. In addition, coherent multiplexing of a remote FPI optical fiber sensor can be used as a point sensor to measure the change of temperature with a random sensing fiber length [34]. For measurement of a multiple state parameter, a full-quartz linear FPI sensor with feedback control of the cavity length is possible to detect acceleration and temperature simultaneously [35].

5.3.3 Sensitivity of an FP-Based Sensor

In recent years, with the development of MEMS technology, it began to play an important role in EFPI ultrasonic sensors. Compared with the IFPI ultrasonic sensor, the EFPI has a small structure and a simple sensing system. Therefore, it is expected for the sensor to be physically placed into the internal space of a transformer-like the transformer winding to perform PD detection. Since the sensing probe of a Fabry–Perot interference structure is prepared additionally, preparation methods and materials of an FP cavity membrane need to be improved and optimized to achieve high sensitivity. However, due to the complex process and urgent requirements for a laser light source and operation technique, the problem of cost is still an important factor restricting further development of a higher sensitivity Fabry–Perot interference structure.

5.3.3.1 The Diaphragm Thickness

Research on EFPI fiber ultrasonic sensors has been reported frequently to optimize the design of the diaphragm size to improve the sensitivity of the sensor [36–38]. Theoretical analysis shows that the natural frequency of the EFPI fiber ultrasonic sensor diaphragm is proportional to its thickness and inversely proportional to the square of the effective radius of the diaphragm. When the natural frequency of the EFPI sensor is fixed, the thinner the diaphragm, the greater the response displacement of the center of the diaphragm. That is, reducing the thickness of the diaphragm is the first measure to improve the sensitivity of the EFPI sensor. For example, a glass tube is spliced to the end of a single-mode fiber and a bubble wall with a thickness of micrometers is prepared at the fiber end through multiple weak discharges and slow pressure release to form an FP cavity, where its sensitivity is increased. This is shown in Figure 5.18 [39]. The outer surface of the FP cavity membrane can also be processed by *fs*-level laser technology, because its sensitivity of stress detection changes proportionally to the membrane

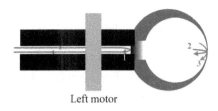

Left motor

Figure 5.18 FP cavity fabrication process of an all-fiber FPI based on an air bubble. Adapted from Yan et al. [39].

thickness. It fabricates a diaphragm-based miniature fiber optic FPI pressure sensor by *fs* laser micromachining [40].

5.3.3.2 The Diaphragm Material

The material properties of the membrane have an impact on detection sensitivity. The material for making the FP cavity diaphragm needs to be uniform and dense, with good wear resistance, reflectivity, hardness, and corrosion resistance. Common materials that can be used for diaphragm preparation include steel, cobalt, tin, gold, silver, platinum, and multiple alloys. However, because of the high cost of gold and platinum, it is rarely used in industry and the current mainstream diaphragm preparation materials include copper, aluminum, and silver.

The reflectivity of the aluminum thin diaphragm is low in the visible light band, but the aluminum diaphragm has high reflectivity in the ultraviolet and infrared bands. Therefore, the aluminum film can be used as a reflective diaphragm in the band below 400 nm, but its wear resistance is poor and it easily reacts with weak acids. The reflectivity of the copper diaphragm at the long wavelength band (600–800 nm) of visible light is 20%–40%. In the short wavelength band (below 600 nm), it is less than 20% and the reflectivity increases significantly above 800 nm. Due to the bright color of the copper diaphragm and the reflectivity at different wavelengths, it can be used as a decorative diaphragm and a thermal reaction diaphragm. The silver diaphragm has a high reflectivity in the visible light range and the infrared band, but the reflectance in the ultraviolet band is low. Considering the high reflectance in the infrared band, high chemical and thermal stability of the silver diaphragm, and the mature silver plating technology, silver is the most common material used for FPI sensors. However, the metal material is not a suitable choice for application in a power transformer.

Some new materials are gradually being used to prepare FP sensing probes, such as graphene, polyethylene glycol ester (PET) [41], biocompatible photonic materials [42], and silicon [43], especially a quartz sleeve [44]. Maybe these are the candidates for EFPI sensors in a power transformer.

5.3.3.3 The Diaphragm Shape

The shape of FPI sensor also affects the frequency response and the effect of detection. The response of FPI is linear when all of the conditions are ideal and it is sensitive to a detection signal. Thus, according to elasticity theory, the center deflection of the FP diaphragm can be expressed as [45]

$$y(P) = 3(1 - u^{1/2})a^4 P/(16Eh^3) \tag{5.2}$$

where P is pressure, h represents diaphragm thickness, a is the radius of extrinsic FPI, E is Young's modulus, u is Poisson's ratio, and y expresses the center deflection of the FP diaphragm. The response frequency is shown as

$$f = 70.2hE^{1/2}(12 - 12u^2)^{-1/2}/\pi a^2 \omega^{1/2} \tag{5.3}$$

where ω is the density of the diaphragm.

The frequency increases with a thick FP diaphragm. Furthermore, the frequency of the circular sensor is higher when the square diaphragm side length is equal to the circular diaphragm diameter and their thicknesses are the same. Generally, the velocity of the circular diaphragm frequency is faster than that of one with a square shape, and it has been demonstrated that the circular sensor has a better performance on ultrasonic detection.

5.3.4 Merits and Drawbacks

There are merits for FP interference to detect PD in a power transformer.

- **Flexible installation.** It is easy to arrange an installation inside the power equipment without affecting normal operation. At the same time, the interference position is located in an air FP cavity, which results in stable polarization, high accuracy, and little noise interference.
- **Simple system.** The acoustic wave in the transformer oil can be detected using an optical fiber acoustic sensor. This sensor is made by bonding silica tubing and a silica diaphragm together to form a sealed fiber optic extrinsic Fabry–Perot interferometer. The diaphragm of the Fabry–Perot interference structure requires fewer devices and the sensing probe is composed of an FP cavity with a small volume.
- **Low cost.** The composition of the FP-based sensor is very simple and the manufacturing cost of FP probes will be cheap once the parameters are determined.

However, some tasks need to be completed to resolve the challenges of FP sensors applied in a power transformer for the long term.

- **Cross interference.** The acoustic wave induces a dynamic pressure on the diaphragm, which leads to vibration of the diaphragm. Therefore, it is very important to design the sensor head to ensure a high enough frequency response and sensitivity to achieve optimum detection of PDs [46].
- **Parameters to be determined.** The size, materials of the diaphragm, and the length of the cavity are parameters that must be considered to guarantee the sensitivity of detection [10]. In addition, FPI sensors with the disadvantage of inherent susceptibility to signal drift need to be further researched.
- **Sing point detection.** The weak vibration caused by PD ultrasonic signal can be effectively perceived only near the FP cavity. In addition, the effectiveness depends on the direction, which means that the sensor angle needs to change with the PD position. To solve this problem, more FPI sensors need to be installed, which makes it a high cost item. On the other hand, it can only cover part of the frequency of the PD acoustic signal due to the narrow bandwidth of the diaphragm sensor. Thus FPI is not suitable for distributed and long-distance detection of PD in power equipment.

5.4 Dual-Beam Interference-Based PD Detection

5.4.1 Principle of Different Interference Structures

In a long-term operation of a high-voltage apparatus, there is always a threat from over-voltage or insulation failure, which can lead to an insulation breakdown phenomenon. During the PD cycle, the discharge energy is released and intensity of the electric field decreases, leading to shrinkage and vibration of defects (like voids) that generate an ultrasonic signal [47, 48]. In addition, under the simultaneous influence of low voltage and a pulsed electric field force, the defect exponentially decays and oscillates in the insulation oil, leading to ultrasonic signals [49]. Apart from an FP-based sensor, some optical fiber sensors can also be applied for vibration sensing. Especially, typical dual-beam interference structures include Mach-Zehnder, Michelson, and Sagnac. Ultrasonic signals propagate in the transformer and cause weak deformation of the optical structure sensing part. At the same time, the refractive index of the optical interference structure is changed due to the elasto-optical effect, which results in a phase difference in the optical signal and is eventually reflected as output light intensity fluctuations [50].

5.4.1.1 Mach-Zehnder Interference

The Mach-Zehnder (M-Z) interference structure, which is different from that of Fabry and Perot, does not require the preparation of optical components, and the system can be built by combining existing optical components. The M-Z interference system mainly includes a laser light source, coupler, reference fiber, sensing fiber, photodetector, and output signal module. As shown in Figure 5.19, 2 × 2 Coupler 1 divides the light from the laser light source into two equal beams with the same direction and wavelength, which go through a reference fiber and sensing fiber to arrive at Coupler 2. Then, the two light beams cause interference in Coupler 2. Once there is an insulation defect in the electric equipment, the ultrasonic signal generated by PD stresses on the surface of the sensing fiber causes optical fiber deformation and gives rise to the change in phase difference in the interference signal.

Based on the dual-beam interference principle and response frequency of a photodetector, the relationship between the output light intensity and optical path difference can be expressed as [51]

$$I = A + E_1 E_2 \cos[2\pi(L + \Delta L)/\lambda] \tag{5.4}$$

In the equation, A is a constant, λ represents the laser source wavelength and E_1, E_2 are the amplitudes of reference fiber and sensing fiber, respectively. L is the initial optical

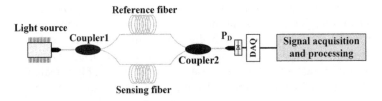

Figure 5.19 Typical structure of the Mach-Zehnder interference system.

path difference and ΔL is the change of optical path difference caused by a PD ultrasonic signal. Equation (5.4) can reflect the relationship between light intensity and optical path difference caused by external stress changes. What is more, the optical path difference can be calculated by recording the light intensity, and the original PD amplitude can be derived.

The M-Z interference with a simple structure, which does not require additional optical components, can detect weak external acoustic signals and its sensitivity can be improved by adding a length of optic fiber. At the same time, due to the absence of reflection in optical signal propagation, the reflected light and optical signal caused by Rayleigh backscattering have different directions with respect to the optical signal emitted by the laser source. Thus, optical noise in an optical path has little effect on the laser signal. However, the sensitivity of the M-Z interference structure is lower than that of traditional UHF sensors and HFCT for PD detection. In an M-Z structure, the length of sensing fiber should be the same as the reference fiber and the reference fiber needs to be completely shielded from the external environment to ensure that it is free from external interference. This brings in the problems of uniformity in length between the sensing fiber and the reference fiber, the isolation of the reference fiber, and a reasonable arrangement after the isolation needs to be considered. Furthermore, it is necessary to continuously adjust the reference arm phase to ensure that wo arms are in an orthogonal state and to obtain the high detection amplitude. However, the PD ultrasonic signal is irregular, its frequency and amplitudes are distributed randomly, and adjustment of the phase cannot be guaranteed to be effective in real time.

5.4.1.2 Michelson Interference

The Michelson interference structure is similar to that of the M-Z structure as they both have reference and sensing fibers. However, the components of the Michelson interference structure have two additional Faraday rotating mirrors, which play a role in reflection. The narrow-band optical signal emitted by the laser light source is divided into two optical signals with the same propagation direction and frequency. They propagate along the reference fiber and sensing fiber, respectively, and are reflected by a Faraday rotating mirror to a previous path. Similarly, vibration of the PD acts on the surface of the sensing fiber, which causes the fiber to be deformed. Due to the influence of the elastooptical effect, the light refractive index changes in the sensing fiber. Furthermore, the phase difference exists in the reference and delay fiber, which leads to the change in light intensity. A typical Michelson optical structure is shown in Figure 5.20.

Figure 5.20 Typical structure of the Michelson interference system.

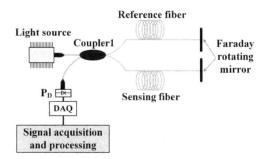

Light intensities reflected by the two Faraday rotating mirrors are expressed as I_1 and I_2, respectively, and the total light intensity after interference can be expressed as [52]

$$I = I_1 + I_2 + 2\sqrt{I_1 I_2} \cos \varphi \tag{5.5}$$

where φ is the phase difference between two beams of light signals, L indicates the length of the sensing fiber and β represents the transmission constant. Thus the change of phase difference is

$$\Delta\varphi = \beta\Delta L + L\Delta\beta. \tag{5.6}$$

Equations (5.5, 5.6) reflect the impact of vibration signals on the length of the sensing fiber and the transmission constant. Finally, the interference light intensity changes. It can be demonstrated that real-time detection of output light intensity can determine whether the device is in normal operation.

The most prominent advantage of the Michelson interference structure is about application of the Faraday rotating mirror, which can eliminate the backscattered light while ensuring polarization state consistency of the two reflected light signals. In this way, the transmission loss of the optical signal does not decay at the reflection process [53]. However, the Michelson interference structure needs to be considered as the consistency of two fiber transmission paths and the isolation and layout of the reference fiber. When the Michelson structure is used for PD detection of power equipment, it is similar to an acoustic emission probe for a multi-path propagation phenomenon of ultrasonic signals inside the power equipment. Meanwhile, the detection amplitude of the Michelson structure performs better in the low frequency band and the vibration signal at normal operation of the power equipment is prone to causing a misjudging phenomenon.

5.4.1.3 Sagnac Interference

The typical application of a Sagnac interference structure is for an industrial gyroscope, which uses an external source to extend the delay of an optical fiber and enhance sensitivity. The Sagnac interferometer-based fiber-optic gyroscope (FOG) was first proposed by Vali and Shorthill in 1976 [54] and was used in the control systems of aircraft stabilization to achieve inertial navigation. The Sagnac interferometer is also applied in current detection and dynamic acoustic detection apart from the well-known rotation rate. When it comes to detecting an ultrasonic signal by Sagnac, it imposes constraints on packaging and detection coil geometries to minimize the effects of magnetic fields, acoustic fields, and thermal transits.

The phase detection can be approached in two ways; one is by open loop and the other is by closed loop, where closed loop is based on open loop and their principles are similar. The Sagnac interference structure includes a light source, coupler, sensing fiber, photodetector, and signal processing module. Due to the reciprocity of the Sagnac optical path, the phase noise of light source will not be converted into system intensity noise. Therefore, the light source of the Sagnac structure is different from interference structures introduced as above, and the Sagnac interferometer is used as a rotation measurement for both civilian and military applications. A broadband light source with a low coherence length helps to reduce the system cost instead of a laser light source [55], as shown in Figure 5.21, where l_1 and l_2 are single-mode optical fibers, which mainly play the role of a transmission optical path and l is a sensing probe surrounded by optical fibers and is used to sense changes of external acoustic waves. The coupler divides

Figure 5.21 Principle detected by a Sagnac interference structure. Source: Modified from Cho et al. [55].

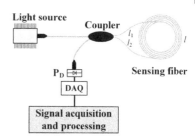

the low-coherence light emitted by the broadband light source into two beams with the same frequency that travel clockwise and counterclockwise. One beam of light is transmitted to the coupler through $l_1 \to l \to l_2$ and the other is propagated through $l_2 \to l \to l_1$. Finally, the two beams meet and interfere at the coupler. Since the propagation paths of the two beams have the same parameter except for direction, the optical path difference between them is zero when the Sagnac interference structure is not affected by any external forces. On the other hand, due to the different times for the two beams to reach the disturbance position when the ultrasonic signal existed, there is an optical path difference between the two light beams, which further causes the interference signal strength to change. The interference signal intensity is detected by demodulation and can be used to obtain the phase difference between the two optical signals, in order to detect the PD ultrasonic signal.

The process of frequency variation is as follows. When the ultrasonic signal is applied to the sensing optical fiber probe l, the effect of the ultrasonic waves on the two arms l_1 and l_2 of the Sagnac interference structure is negligible. The optical field intensities E_L and E_R of the two beams reaching the photodetector can be expressed as

$$E_L = A \exp j[\omega t - \varphi_s(t - \tau_L) + \varphi_1] \tag{5.7}$$

$$E_R = A \exp j[\omega t - \varphi_s(t - \tau_R) + \varphi_2] \tag{5.8}$$

where A is a coefficient that is proportional to the input light amplitude and the input loss of coupler, ω is light wave frequency, φ_s is the phase change of two beams in the sensing optical fiber probe due to the ultrasonic signal, τ_L and τ_R are the propagation times of the two beams propagating in a clockwise direction and counterclockwise direction, respectively, and φ_1 and φ_2 are initial phases of two beams in a sensing optical fiber probe, which are related to the position of the ultrasonic wave on the probe. From Eqs. (5.7, 5.8), the intensity of the output light to the photodetector is expressed as

$$I_{out} \propto (E_L + E_R) \cdot (E_L + E_R)^H = 2A^2[1 + \cos(\Delta\varphi_s + \Delta\varphi)] \tag{5.9}$$

In Eq. (5.9),

$$\Delta\varphi = \varphi_1 - \varphi_2 \tag{5.10}$$

$$\Delta\varphi_s = \varphi_s(t - \tau_R) - \varphi_s(t - \tau_L). \tag{5.11}$$

Since the advantages of a Sagnac interferometer, such as small size, anti-EMI, and safety, have been found, it has been widely used in offshore oil exploration, intelligent oil wells, anti-submarine warfare, port traffic monitoring, seismic monitoring, aging monitoring of civil buildings, as well as for PD detection [56]. The optical fiber sensor array can not only find the PD disturbance but it can also locate the PD source position in real time. Thus, the Sagnac interference sensor has been widely researched in recent years.

At the same time, there are many variations of the Sagnac interference structure, and a depolarizer can be used to solve the polarization decline problem, thereby reducing the complexity of the system greatly [57]. On the other hand, when it is used for PD ultrasonic signal detection, the frequency response characteristics of the Sagnac structure can effectively suppress the low frequency background noise of substations to improve the signal-to-noise ratio of detection. In addition, its sensing probe is wound directly by the section of optical fiber and therefore there is no need for the reference arm, which reduces the difficulty of system construction and facilitates distributed measurement. These characteristics bring advantages to PD detection based on the Sagnac interference structure in a power system. However, due to the Sagnac interference circular structure, it is severely restricted by its placement.

5.4.2 Application Cases

5.4.2.1 PD Detection Based on Mach-Zehnder

The experimental setup for the detection of PDs based on the Mach-Zehnder interferometer has been verified in laboratory conditions [58]. As to the configuration and detail, the coherent light source used is an He–Ne laser that operates at a wavelength of 633 nm in the visible range. This source also facilitates the alignment of the optical arrangement. Two fiber arms are composed of the same length of optical fiber using two identical fiber coils, which helps to prevent errors due to optical path differences. The coil of the sensing arm of the interferometer is submerged in transformer oil, which is exposed to perturbations caused by PD acoustic waves. The other arm of the interferometer is isolated from the impact of the acoustic wave, being used as the reference optical path, as displayed in Figure 5.22.

In order to cover the frequency range of PD-induced acoustic emission, bend-insensitive fiber is recommended to be employed for implementation of the fiber sensing coil.

5.4.2.2 PD Detection Based on Michelson

To achieve non-invasive detection in a power equipment, an optical sensing probe is proposed as in Figure 5.23 [59]. It is wound around the electrical equipment surface

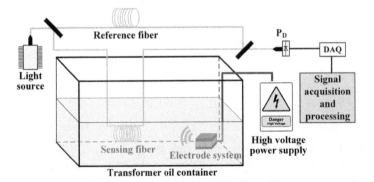

Figure 5.22 Mach-Zehnder interference is used to detect PD in transformer oil.

Figure 5.23 The fiber winding on the surface of electrical equipment. Source: Modified from Zargari and Blackbum [59].

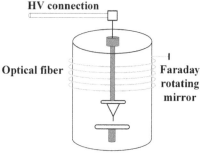

Figure 5.24 Schematic diagram of a sensing probe with skeleton. Source: Modified from Zhang et al. [60].

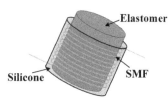

merely through a single-mode optical fiber. In this manner, as the length of the sensing probe increases, its sensitivity could be enhanced. Moreover, it indicates that the sensor can be wound and installed in electrical equipment to detect PD as conveniently as possible.

An optical fiber is also wound on a cylinder with a great inner diameter for detecting PD in transformer oil directly [58]. The size of the cylinder is a key parameter, as too large a size of the sensing fiber is impossible to use for the modulation. However, a more exquisite and smaller sensing Michelson probe can be helpful, as shown in Figure 5.24 [60]. To guarantee the detection sensitivity, the elastic cylinder with a high Poisson ratio is chosen as a skeleton and the single-mode optical fiber is wound around the skeleton. Moreover, the sensor probe can be fixed on anywhere for actual detection and the sensitivity is ensured at the same time.

5.4.2.3 PD Detection Based on Sagnac

The Sagnac optical structure of PD detection is also useful for ultrasonic detection. The main sensing part of the Sagnac interference structure is a closed optical coil. If there is no vibration caused by ultrasonic signals outside, no phase difference occurs between the clockwise and counterclockwise light. On the contrary, once the vibration caused by PD stresses occur on the surface of the fiber, the refractive index (RI) fluctuation acts in the fiber core, which eventually changes the intensity of the output light. Then, the amplitude and frequency of the PD ultrasonic signal can be reflected through demodulation of the output light intensity, as shown in Figure 5.25.

A PD model is built to illustrate the availability of optical interference structures in PD detection. As shown in Figure 5.26 [56], the needle plate discharge model is chosen and air is filled with the interval between the needle and mental plate. The needle electrode is connected to a high voltage supply and the plate electrode is grounded. Taking into account the reliability requirements, optical sensors, AE sensors, and HFCT sensors are installed simultaneously to detect PD.

The detection waveform of the optical structure is shown in Figure 5.27, which accords with the results of conventional AE and HFCT sensors. The results demonstrated that

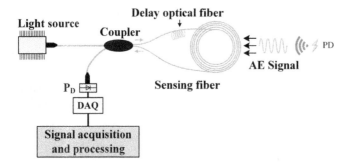

Figure 5.25 Sagnac interference structure to detect PD ultrasonic signals.

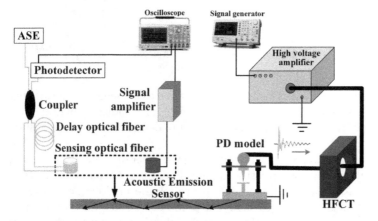

Figure 5.26 Experimental platform for ultrasonic detection of real PD. Source: Jiang et al. [56].

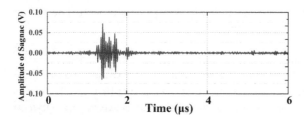

Figure 5.27 The actual PD signals received by the optical fiber sensing probe.

optical fiber is also effective in actual PD detection and can even be used to explore the PD ultrasonic propagation characteristics.

However, PD signals are often intermittent in actual detection and the statistics of ultrasonic amplitudes are difficult, where sinusoidal signals with variable frequencies excited by the AE sensor (R3α) are used to replace the PD model. At the same frequency, amplitudes of different positions at the same time are recorded. Furthermore, the attenuation characteristics of the ultrasonic signals of the Sagnac sensor are compared with the AE sensor and simulations at different frequencies, as shown in Figure 5.28.

A large amount of refraction, reflection, and absorption occurs on the surface of a transformer wall, resulting in a partial energy loss [56]. Cumulative energy per unit area of wall is inversely proportional to distance. Therefore, the far propagation distance makes the great attenuation on the amplitude of the received ultrasonic signal. At the

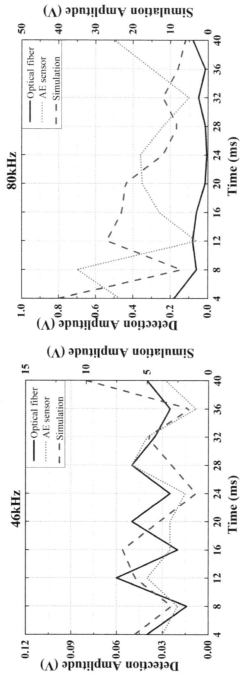

Figure 5.28 Typical Sagnac detection and simulation amplitudes at 46 and 80 kHz.

same time, the characteristics of the ultrasonic signal amplitude with distance in the shell is not monotonously decreasing. Due to the interference in ultrasonic propagation, the wave superposition causes the maximum amplitude only when the vibration direction, frequency, and phase of ultrasound waves are the same. Similarly, there are multiple minimum amplitude points because of wave cancelation. This means that the amplitude of the ultrasonic wave is fluctuating and decreases with the distance parameter.

5.4.3 Sensitivity of an Interference-Based Sensor

5.4.3.1 Sensor Parameter Variation

In the practical application of power equipment detection, PD ultrasonic signals are prone to attenuate when they propagate through various components in a high-voltage equipment. It undoubtedly leads to low amplitude of the signal received by the optical sensing probe, even if it is unable to receive the signal at all. Hence the sensitivity enhancement of an optical sensor is a very important issue to make it effective.

In the Mach-Zehner, Michelson, and Sagnac interference structures, the detection principle is quite similar and the Sagnac interference structure is taken as an example to analyze the factors such as length of sensing fiber and winding forms.

In the preparation processing of interference structures, an optical fiber with length much longer than the sensing fiber is often inserted between the sensing probe and the coupler, which is called a delay fiber as it acts as a phase delay. Suppose the length of the sensing fiber is L_s, the delay fiber length is L_t, and the total length of the fiber is L. In addition, λ_0 is the average wavelength of the optical signal emitted by the light source and n represents the refractive index of the fiber core; β is the propagation coefficient. Then the phase delay of the output light intensity through the Sagnac interference structure is

$$\varphi = {2\pi n L}/{\lambda_0} = \beta L. \tag{5.12}$$

When the fiber surface is subject to mechanical stress, the length of the sensing fiber changes slightly with the strain. At this time, the elasto-optical effect and Poisson effect make an impact on the refractive index and diameter of the fiber core [61, 62]. However, due to the tiny diameter change of the fiber core, it can be ignored. Thus, the phase difference of the optical signal caused by an ultrasonic wave can be expressed as

$$\Delta\varphi = \beta\Delta L + L\Delta\beta = \beta\Delta L + L\frac{\partial\beta}{\partial n}\Delta n. \tag{5.13}$$

The mechanical force on the optical fiber is usually regarded as static pressure. Assume that there is only positive strain on the optical fiber surface; then the phase difference caused by the strain effect is

$$\Delta\varphi\varepsilon = \beta\Delta L = -\frac{L\beta P(1-\mu)}{E} \tag{5.14}$$

where P is the sound pressure, and μ and E represent Poisson's ratio and the Young modulus, respectively. When the Poisson effect is neglected, the influence of the propagation coefficient on the phase difference is shown as

$$\Delta\varphi_s = -\frac{Lk0n^3 P(1-2\mu)(p_{11}+2p_{12})}{2E} \tag{5.15}$$

where p_{11} and p_{12} represent the fiber's anisotropic optical coefficients, respectively. When the phase difference caused by the strain effect and the elasto-optical effect is substituted into Eq. (5.), the equation of sound pressure sensitivity is

$$\frac{\delta\varphi}{P_0} = 2\beta M_0 Ls \sin\left(\frac{\omega n}{2c}L_t\right) \tag{5.16}$$

where P_0 shows the pressure amplitude of the ultrasonic signal acting on the optical fiber surface, ω indicates angular frequency, M_0 is a constant, and c represents the speed of light propagation in the optical fiber. According to Eq. (5.16), the sensitivity of interference structure changes periodically with the delay fiber length and monotonically increases with the length of the sensing fiber when the external signal is constant and the external interference signals are the same.

In addition, the winding radius of the sensing fiber will influence the detection effect. Normally, the diameter of the sensing fiber winding needs to be as small as possible in order to capture the signal in the ultrasonic frequency range generated by PD. That means its diameter is required to be much smaller than the wavelength of the ultrasonic signal to ensure that the sensing fiber curves when the external stress occurs. However, a small winding diameter is easy to increase the bending loss, as shown in Figure 5.29. Therefore, the winding radius of the sensing fiber should be weighted to guarantee a high detection signal-to-noise ratio and sensitivity.

There is a common requirement for long sensing fibers for Mach-Zehnder, Michelson, and Sagnac interference structures and the sensing fibers are generally wound as a ring. However, the winding method and skeleton material of the winding process also have an impact on the detection sensitivity of the interference system. At present, the common sensing fiber winding methods include planar radial winding [63, 64] and solid axial winding [65, 66].

The planar radial winding method consists of winding the sensing optical fiber tightly around the surface on the device or solid disk, and the overall distribution is flat, as

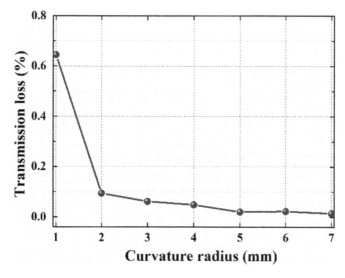

Figure 5.29 Variation trend of output loss with bending radius.

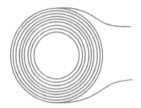

Figure 5.30 Planar radial winding method.

shown in Figure 5.30. The planar radial winding structure is simple and does not need complicated processes. Also, the large contact area between the sensing probe and the winding arrangement causes only a small transmission loss of optical signal. However, the fiber ring diameter winding by this method is inconsistent in a wide frequency band. It is easy for the output signal to have a zero response during the modulation of the sensing fiber by the ultrasonic signal. In addition, the plane of the fiber is a single-layer spiral structure with poor stability, which can easily be bent, loose, and damaged, which affects the operation. Therefore, the sensing probe of the planar radial winding method is only installed on the surface of a specific device.

The sensing probe of the solid axial winding method is superimposed within a small area. Furthermore, due to the existence of the sealing layer and skeleton, the sensing probe is not damaged by an external force to any extent and its stability is improved with the tight and solid package. It lays the foundation for achieving the flexible placement goal of an optical interference structure. However, the fiber with soft materials is easily damaged and difficult to shape. To ensure close contact between fibers and increase the contact area between the device and sensing probe, it is usually wound and fixed by means of a skeleton or mandrel.

5.4.3.2 Phase Modulation and Demodulation Techniques

The effect of stress variations on interference structure is naturally the phase modulation process. When a sinusoidal ultrasonic signal with a known frequency is applied to a sensing fiber and the DC signal is ignored, the output light intensity is as shown below:

$$I_{out} \propto \cos\left[\gamma \sin \omega_u \left(t - \frac{\tau}{2}\right) + \varphi_0\right].$$

(5.17)

In the equation, $\gamma = 2\varphi_{s0} \sin(\omega_u \tau/2)$, ω_u is frequency of ultrasonic wave, φ_{s0} represents the change amplitude of phase, which is proportional to the intensity of ultrasonic wave, $\tau \approx Ld/c$ is the propagation time of light in the fiber, L_d is the length of delayed fiber, c shows the propagation speed of light in the fiber, about 2×10^8 m/s, and φ_0 is the initial phase shift. It can be seen that the initial phase has an influence on the intensity of the output light. If the initial phase is located at the peak or trough of the sinusoidal signal, visibility of the interference signal decreases and the output light intensity reduces. Therefore, the initial phase needs to be adjusted to ensure detection sensitivity. In addition, the frequency of the output light intensity is doubled by the modulation when the input signal is a sinusoidal signal and the waveform becomes complex.

Apart from a Faraday rotating mirror, the optical elements that enhance the detection amplitude of the interference optical structure also include the phase modulator [67, 68]. The phase modulator mainly improves the detection amplitude and

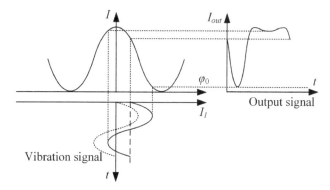

Figure 5.31 The effect of phase modulation on the output light intensity when the initial phase is arbitrary.

signal-to-noise ratio of the system by changing the initial phase. The mechanism of its effect on the interference structure is shown in Figure 5.31. When the phase modulator is absent, the small initial phase leads the amplitude of the modulated output signal because the external ultrasonic signal is too weak to be submerged. It is illustrated that the background noise of the field has a great influence in detection sensitivity. Moreover, the initial phase shift is shifted toward $\pi/2$ after the phase modulator is employed, and the amplitude of the output signal increases simultaneously. When the phase shifts at $k\pi + \pi/2$, $k = 0, 1, 2, 3, \ldots$, the response value reaches the maximum. Based on the above analysis, the phase modulator helps to improve detection accuracy.

In order to eliminate or reduce the influence of the initial phase on output light intensity and reflect the real frequency response, demodulation technology is adopted to obtain the original signal. The existing demodulation algorithms can be partitioned into the homodyne demodulation method [69, 70] and the heterodyne demodulation method [71–73].

The homodyne demodulation method mainly includes an active homodyne method, a phase generation carrier (PGC) method, and a 3×3 coupler method. The active homodyne method [74] has a long history but is a simple process. By adding feedback to the interference structure, the initial phase always works near the operating point and gives rise to an output signal fluctuation. However, due to the existence of feedback, the operating frequency band and dynamic range are limited, which increases the design complexity of the interference structure.

The PGC method is a passive homodyne method and usually introduces a large amplitude phase modulation outside the frequency band of the PD ultrasonic signal. It has the advantages of a large dynamic range, high sensitivity, and detection accuracy. In addition, the common PGC methods can combine with a mathematical tool, such as the differential cross-multiplication (DCM) method [61, 75] and the Arc tangent (Arctan) method [76–78], which are shown in Figure 5.32.

The PGC-DCM method is severely affected by fluctuation of light intensity during the demodulation process. When there is a strong light intensity interference, the demodulation result is distorted. In addition, the depth of phase modulation also affects the demodulation result and excellent phase modulation depth needs to be considered to obtain the best demodulation effect. The PGC-Arc tangent method makes up

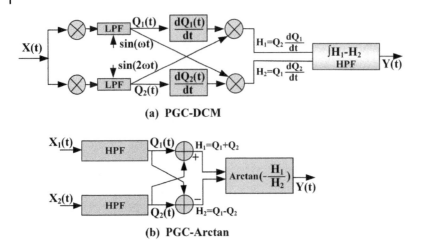

(a) PGC-DCM

(b) PGC-Arctan

Figure 5.32 The demodulation principle of phase generation carriers.

for the shortcomings of DCM about light intensity fluctuations, and improves the demodulation stability of the output signal. However, the effect of the phase modulation depth on demodulation stability and gain is still evident.

At the same time, multiple uncertain factors are introduced to the demodulation process of the PGC method, such as the visibility of interference structure phase and the difference between the double frequency of the vibration signal and carrier signal. To solve the problem of uncertain factors, there are many derivative PGC demodulation methods [79–81], like the asymmetric division algorithm based on fundamental frequency mixing, the arc tangent-differential and self-multiplication algorithm, etc., which increase additional complexity of the process.

The 3×3 coupler method [82–84] is a passive homodyne method with the help of three output ports with a 3×3 coupler (120° phase difference, respectively). After two signals processed by a high-pass filter and addition, a set of mutually orthogonal signals can be obtained. Furthermore, two mutually orthogonal signals are divided and become negative and the vibration signal can be demodulated by dividing it by $\sqrt{3}$ and performing an arc tangent transformation. Compared with PGC, the principle of the 3×3 coupler method is simpler and the phase demodulation range is greatly increased. However, the 3×3 coupler, as the most important device in the topology, cannot be ideal in reality. That means there is a deviation between the phase difference of the output ports and 120°, which can easily cause unsatisfactory detection results.

Heterodyne demodulation is an algorithm in which the signal is modulated to a high frequency band, making the signal less susceptible from interference of low frequency background noise. Moreover, the detection accuracy of the interference structure is improved and the demodulation performance is superior to the PGC because the heterodyne demodulation method is insensitive to the polarization state. However, due to the necessity of frequency-shifting devices in signal processing, the cost of demodulation is greatly increased. Therefore, heterodyne demodulation is often limited by cost in actual optical interference structure detection.

5.4.4 Merits and Drawbacks

Compared with the Fabry–Perot interference, dual-beam interference with three typical structures have some unique merits.

- **All-fiber structure.** The dual-beam interference adopts an all-fiber structure and the sensing part is directly wound by an optical fiber. There is no need to prepare additional sensing probes or membranes.
- **Flexible measurement.** Owing to its all-fiber structure, it can realize the distributed measurement of PD in a large power transformer and immunity to electromagnetic forces enables inside installation.
- **Reasonable cost.** With the employment of broadband light modules and universal optic fibers, the cost of the measurement system is relatively reasonable and acceptable.

As to the challenges of dual-beam type sensors, there are mainly issues on the performance.

- **Detection sensitivity.** For the sensitivity problem in PD detection, the measurement accuracy can be improved by increasing or decreasing the number of sensing probe windings or delay the fiber length. It is also necessary to ensure the stable operation of dual-beam interference at the working point to improve the detection sensitivity.
- **Probe fabrication.** It is difficult to assume that the environment of the two optical fibers is completely consistent and the reference arm is strictly stable, since two optical fibers are used as the sensing arm and the reference arm.
- **Response frequency.** The suitable frequency of PD detection for specific occasions are different, and the bandwidth and frequency response of interferometric sensors are hard to be controlled. In addition, it is necessary to reduce optical noise and temperature noise in oder to maintain overall structural stability and positioning accuracy.

5.5 Multiplexing Technology of an Optical Sensor

5.5.1 Multiplexing Technique with the Same Structure

The single optical principle has its own shortcomings and cannot meet the actual needs of detection at times. In order to achieve efficient and accurate measurement, some composite structures composed of different optical units are used for actual PD detection. The combination of FBG and interference can be seen as a new interferometer where the sensing probe integrates with the FBG as the reflector. In addition, the combination with similar interferences or different principles makes it possible to achieve multi-channel measurements in actual detection. It is called the multiplexing technique of optical fiber sensing units.

According to the structure of the multiplexing technology, it can be partitioned into space division multiplexing (SDM), TDM, wavelength division multiplexing (WDM) (also called frequency division multiplexing), and hybrid multiplexing. The obvious feature of SDM is that each FBG sensor works independently, there is no interference between each optical structure, and it has a high signal-to-noise ratio. From another perspective, if the sensors are connected in series, optical fiber breakage causes breakdown of the entire detection system. Therefore, the parallel structure between sensors of SDM effectively avoids the impact of a sensor failure on the system. However, the SDM sensor has a significant disadvantage in that the utilization rate of the light source is poor. TDM is composed of many sensor arrays in each channel and there is a delay fiber among each channel. The pulse signal generator controls the drive of the light source. The delay time corresponds to the delay optical fiber. When it comes to WDM, the frequency band interval of the light source, the utilization rate of the spectrum, and the transmission distance need to be considered.

Owing to rapid optical fiber development during the past few years, it has made great progress. However, considering the weak strain caused by an actual PD signal and the presence of multiple noise signals in substations, the reliability and stability of optical sensors are difficult to meet requirements in practical applications. Designing high performance optical fiber sensors is still the challenging research for an optical fiber sensor system. The use of multiple multiplexing technologies with the same basic structure helps to reduce optical noise while reliably detecting a PD signal in real time.

- **Dual M-Z interferometer.** In order to make the M-Z interferometer system accurately locate an external ultrasonic signal of PD, a dual M-Z interferometric fiber sensor is proposed [85], which is shown in Figure 5.33. The light emitted by a light source is divided into two beams of clockwise transmission and counterclockwise transmission through coupler C1. The light transmitted along clockwise direction is then divided again by coupler C2. Two beams of light split at the second time travel through a reference fiber and a sensing fiber, respectively, and the interference occurs at coupler C3. A photodetector detects the interference signal and above paths form the M-Z interferometer. The light transmitted counterclockwise is split into two beams by coupler C3 and passes through the reference fiber and the sensor fiber. Furthermore, they interfere at coupler C2, which is detected by the photodetector. Another M-Z interferometer is constructed to form a dual M-Z interferometric fiber sensor. Compared with the Michelson interferometer, the structure of the dual M-Z interferometric structure with WDM is simple and does not require two light sources. In addition, the dual M-Z interferometer has a better de-noising preference than a

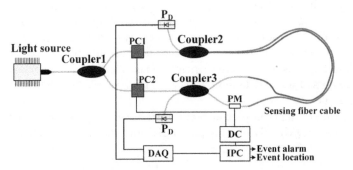

Figure 5.33 A dual M-Z interferometric fiber sensor based on the WDM. Source: Modified from Lamela et al. [85].

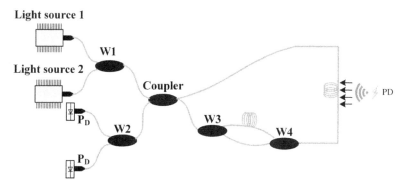

Figure 5.34 Optic fiber sensor of a dual Sagnac interference PD detection system. Source: Modified from Russell et al. [86].

single M-Z sensor and is more suitable for transformer fault detection with serious background noise in the field.

- **Dual Sagnac interferometer.** Since the single Sagnac interferometer adopts a zero frequency point positioning method to locate the external disturbance signal, the output signal is limited in a constant frequency range. Therefore, a dual Sagnac interferometer can be proposed to solve the problem; its structure is shown in Figure 5.34 [86]. Two light sources emit light with different wavelengths and the two beams of light are combined by a wavelength division multiplexer W1 to synthesize a single beam of light. A coupler divides the light into two beams to transmit clockwise and counterclockwise. The light transmitted clockwise is divided by the wavelength division multiplexer W4 into light with different wavelengths after passing through a sensing fiber. Then two beams of light are synthesized by the wavelength division multiplexer W3 and reach a coupler. The counterclockwise light transmit process is similar to that of the clockwise light and travels though W4 and W3 sequentially. Two Sagnac interferometers with different wavelengths employ one sensing fiber to increase sensitivity. The problem of the zero-frequency point positioning method has been solved. In addition, only one optical fiber is adopted to be the sensing fiber in the dual-Sagnac structure. In the PD detection of a large-scale transformer, it helps to simplify the overall layout, while not affecting the normal operation and layout of other sensors. However, the use of two light sources increases the system cost. Furthermore, when the ultrasonic signal is applied to a central position, the phase difference between the two beams is zero and the sensor cannot sense the signal.
- **Multi-FBG/FPI sensors based on SDM, TDM, and WDM.** The FPI sensor is similar to the FBG sensor, as the sensing probe is a point measurement relative to the entire system. Therefore, FPI and FBG sensors can achieve multiple forms of multiplexing, such as SDM, TDM, and WDM.

SDM technology encodes each FBG/FPI sensor with the help of different space positions [87]. Many FBG/FPI sensors are connected in parallel along the optical fiber to form a sensor network. In this way, each sensor has its own independent sensing channel and the SDM optical interference structure uses optical switches to change the channels. Only one channel is selected during demodulation to avoid mutual interference between channels. The SDM structure is the simplest approach. The space division is selected to achieve multiplexing and all interference circuits have detection and receiving devices but share one light source, as shown in Figure 5.35 [88].

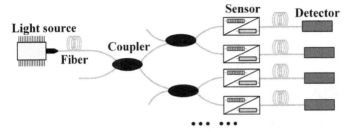

Figure 5.35 Principle and basic structure of space division multiplexing. Source: Modified from Comanici et al. [88].

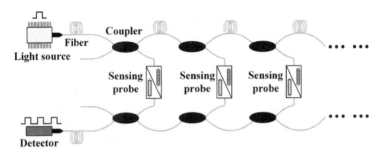

Figure 5.36 Principle and basic structure of time division multiplexing. Source: Modified from Wang et al. [89].

The principle of TDM technology is shown in Figure 5.36 [89], which is composed of several sensor arrays in one channel and there is a delay fiber before each channel. The pulse signal generator controls the driving power of the light source. The delay time corresponds to the delay optical fiber, which is installed before each sensor, and they control each switch of the switch array. At the end of the delay optical fibers, a matched filter demodulator is connected [18, 89].

When the pulse signal generator gives out a pulse, the light source emits a light signal. These lights no longer use wavelength coding, but delay time coding. When the reflected light illuminates a switch array, a high-speed switch array turns on the corresponding switch according to the delay time of the incident light. This method requires that the time interval between two pulses is much larger than the optical delay time. A fast response is hard to achieve when there are too many sensors multiplexed. At the same time, the output optical power of the light source cannot meet the requirements.

Figure 5.37 shows the principle of the WDM technique. The signal fed back by the sensor array is filtered by a tunable filter, and the optical signals are multiplexed by multiple wavelengths, which are demodulated into single wavelength signals one by one. The signal with different wavelengths corresponds to a fixed detection point in the sensors array. By processing the demodulated signal, distributed sensing information can be obtained [90].

The WDM technique mixes two optical signals of different wavelengths carrying phase frequency and other information through the multiplexer. After transmittance of the signal, it is de-multiplexed at a terminal to separate the mixed signal. Moreover, WDM is required to consider the frequency band interval of the light source, the utilization rate of the spectrum, and the transmission distance. The detection device brings in shot noise because it works with optical signals of different wavelengths

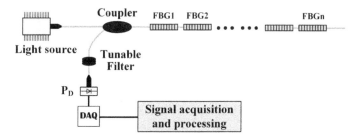

Figure 5.37 Principle and basic structure of wavelength division multiplexing.

simultaneously. Owing to high multiplexing efficiency, the WDM provides a wide detection range and high positioning accuracy to interference structures.

5.5.2 Multiplexing Technique with the Different Structures

Sometimes it is difficult for the single-principle optical multiplexing technology to meet the requirements of multi-parameter detection, so optical structures based on different principles are combined together. The combination of different principles not only benefits in de-noising but also helps to improve the precision of a PD fault location.

- **Multiplexing structure of Sagnac and M-Z.** The structure of a sensor based on Sagnac and M-Z is a typical application of multiplexing technology with a different structure, which is shown in Figure 5.38 [46]. The combination of Sagnac and M-Z can be treated as a linear Sagnac structure. The light emitted by a light source propagates clockwise and counterclockwise, respectively. Among them, the light propagating in the clockwise direction is divided into two channels for propagation, a: Coupler 1 → Delay fiber → Coupler 2 → Optical fiber and b: Coupler 1 → Delay fiber → Coupler 2 → Delay fiber; the light propagating in the counterclockwise direction is also divided into two channels, c: Coupler 1 → Optical fiber → Coupler 2 → Delay fiber and d: Coupler 1 → Optical fiber → Coupler 2 → Fiber. To ensure that the light is transmitted along the above path, it is necessary to make the delay fiber longer than the coherence length of the laser. Due to this condition, only path a and path c meet the interference condition of a zero optical path difference; they just interfere in the optical coupler. The sensitivity of a composite interference structure is higher than most of the others because light travels through the reference fiber twice. In addition, this combined structure colligates the advantages of two sensor structures. Sagnac can detect the action of external PD with high sensitivity and realizes the early warning of the PD event. Like-for-like, the M-Z interferometer also has this potential application.
- **Multiplexing structure of FBG and Michelson.** Since FBG senses changes of strain and temperature by extracting a wavelength shift from reflected light, a wavelength demodulator can measure the center wavelength of the FBG sensor. There are many methods used for wavelength demodulation of FBG sensors. One method utilizes the

Figure 5.38 The structure of sensors combined by Sagnac and M-Z. Source: Modified from Wang et al. [46].

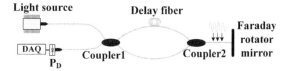

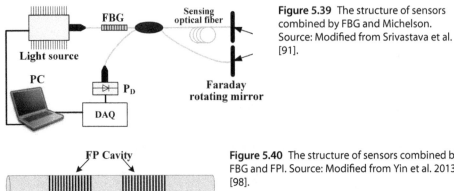

Figure 5.39 The structure of sensors combined by FBG and Michelson. Source: Modified from Srivastava et al. [91].

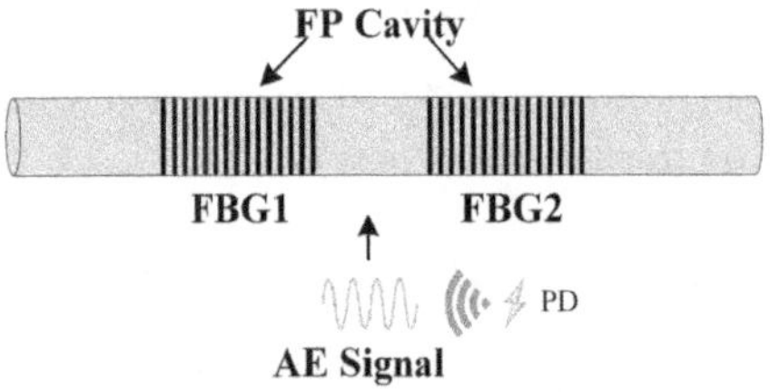

Figure 5.40 The structure of sensors combined by FBG and FPI. Source: Modified from Yin et al. 2013 [98].

Michelson interference structure to demodulate. The Michelson interferometer with an unbalanced optical path difference is able to convert the wavelength shift into an optical phase shift, as shown in Figure 5.39 [91]. It is suitable for demodulation of the dynamic strain sensors based on the Michelson method but requires an active compensation feedback to keep the two interfering beams orthogonal for linear detection of small strains. Without active compensation, the two interfering beams deviate from the optimal operating point due to environmental impact. Therefore, the Michelson method is not suitable for demodulating multiple sensors [92].

In a similar way, the FBG sensor can also be combined with M-Z interference structures [93] and Sagnac structures [94, 95].

- **Multiplexing structure of FBG and FPI.** FBG can also be combined with FPI to form an ultrasonic sensor [96, 97]. The single-mode fiber is engraved with two sections of gratings at the same parameters. FPG in a single mode fiber is regarded as a reflection mirror to compose an FP cavity in an FPI sensor, as shown in Figure 5.40 [98]. When an ultrasonic signal acts on the fiber grating FP cavity, it causes a slight change in the length of the FP cavity and the optical path in the interferometer is modulated. The density of the interference spectrum of it changes rapidly, accordingly the phase of the output light. Then the intensity of the modulated light changes to carry the sensing information of PD. Once the incident light is highly coherent, demodulation can be achieved by measuring the change of light intensity at a specific wavelength within the transmission spectrum; when the incident light has low coherence, the resonance fringe spacing on the reflection spectrum can be measured and a spacing change is used to calculate the corresponding changes of stress. The grating-based FP cavity helps to select the wavelength. This structure has a very high sensitivity and is suitable for detection of high-frequency transient ultrasonic signals, especially for PD measurements.

5.5.3 Distributed Optical Sensing Technique

Distributed optical fiber sensing technology is a new application in PD detection of power system equipment, which has the characteristics of high sensitivity, a large dynamic range, and long-distance continuous monitoring. This kind of sensing technique only requires a light source and an optical fiber to achieve fault detection and transmit on-line monitoring signals, even with the transmission distance at thousands or tens of thousands of meters. In addition, distributed optical sensing is able to carry large amount of information, which helps to reduce the average cost of detection. Thus, distribution detection has become an important topic in the research of optical interference instrumentation. Generally, distributed optical sensing topologies can be categorized as quasi-distributed optical fiber sensors [99, 100] or fully distributed optical fiber sensors [101, 102].

The quasi-distributed fiber optic sensing technique is usually realized by FBG. The FBG with different grating periods is engraved at intervals on the same optical fiber to achieve several distributed PD detections of power equipment [18], just like the multi-point temperature measurement mentioned in Chapter 2. Compared with the fully distributed optical fiber sensor, the quasi-distributed sensor is a distributed structure composed of multiple points, and it cannot realize PD detection with a high resolution in space. Some blind zones are unavoidable.

The fully distributed optical instrumentation adopts an ordinary optical fiber to act as sensing elements. The fully distributed optical fiber sensors can be divided into two categories, one is dual-beams interference including M-Z [103], Michelson [104], and Sagnac [46] interference structures, but with special algorithms and design to detect PD, and the other is the optical time domain reflection (OTDR) technique.

M-Z and Michelson need to isolate the reference fiber, which increases the difficulty of arrangement. The sensing structure using Sagnac as a distributed measurement is shown in Figure 5.41. The PD source can be detected but the annular sensing probe detects the change of physical field symmetrically. The light transmitted clockwise and anticlockwise is affected by the same phase modulation, which leads to counteracted signals. This means that the reciprocal effect of two lights results in the disappearance of the interference signal.

A Sagnac interferometer has an advantage in cost and arrangement. However, it is necessary to isolate half of the optical fiber of the ring structure from the sound field and change it into a non-sensing optical fiber. A linear construction is adopted to solve this problem, as shown in Figure 5.42. This structure uses a Faraday rotating mirror (FRM) to change the ring structure of the Sagnac interferometer into a linear structure, which is suitable for distributed optical detection technology and maintains the zero optical path difference characteristic of the traditional Sagnac interferometer. However, it still

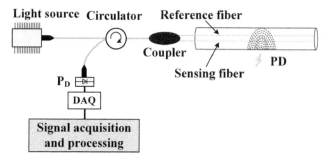

Figure 5.41 PD detection system based on a Sagnac interferometer.

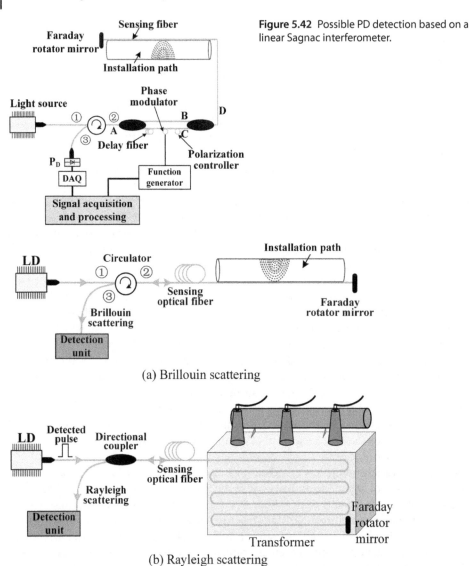

Figure 5.42 Possible PD detection based on a linear Sagnac interferometer.

(a) Brillouin scattering

(b) Rayleigh scattering

Figure 5.43 Layout and installation of an optical time domain reflection technique.

has some problems concerning the polarization state, diffuse reflection, and waveform distortion between the interference beams.

The OTDR technique is based on optical fiber backscattering. Every section in the optical fiber is a sensitive unit, working as both a sensing unit and the signal transmission channel. In theory, the sensing distance is as long as or more than tens of kilometers. Based on the difference of detection signals, OTDR technology can be divided into Brillouin scattering [105–107], Rayleigh scattering [108–110], and Raman scattering. Raman scattering has been explained in the previous chapter and will not be mentioned again here. The layout and installation of Brillouin scattering and Rayleigh scattering are shown in Figure 5.43. The linear layout is used in long-distance detection and the S-shaped layout is used in large-area discharge detection. In addition, due to the structures of Brillouin scattering and Rayleigh scattering being similar, they can be interchanged in an actual installation. This means that both Brillouin and Rayleigh scattering help to detect PD activities at long distances.

When the PD-induced ultrasonic signal propagates in an optical fiber, its pressure difference causes a change in the refractive index of the optic fiber, which affects spontaneous scattering of the transmitting light. This scattering is called Brillouin scattering and propagation of sound brings in the pressure difference and refractive index changes periodically. It also causes the Doppler frequency shift on the scattered light frequency and the ultrasonic signal can be calculated by detecting the frequency shift. The distributed sensing structure of Rayleigh scattering and Brillouin scattering is basically consistent, the only difference being that the measurement of Rayleigh scattering emphasizes the relationship between loss and fiber length.

5.6 Conclusion

PD is one of the key concerns for an insulation system in power transformers to ensure safe and stable operation. The PD induced weak acoustic emission provides an approach to an optic fiber solution with the natural merits of a built-in installation, immunity to EMI, and distributed sensing, etc. Three main optical techniques based on different principles and their combinations are analyzed and included for PD detection in this chapter.

Both a broadband light source and a laser source can be used for FBG PD measurement; the main point is to transduce PD-induced ultrasonic vibration into dynamic strain. Especially, phase shift FBG with a narrow linewidth helps to improve demodulation of the PD measurement.

As an alternative approach, the FPI sensor including typical structures of IFPI and EFPI, also help to sense PD activities. In addition, the FPI cavity is the sensing part of the FPI structure, which can be divided into a thin film structure and a microcavity structure. Highly sensitive PD detection can be achieved by changing the thickness material of the diaphragm FP cavity and the length of the microcavity FP cavity.

The two-beam interference structure for detecting PD based on the principle of optical interference is in an all-fiber manner, mainly including Mach-Zehnder, Michelson, and Sagnac types. They work as optical-fiber ultrasonic sensors, which are designed to be sensitive to acoustic signals generated by PD. The key is the sensing probe and its detection sensitivity can be improved by changing the winding area and fiber length or by adding a phase modulator. In addition, adding a delay fiber and changing its length also help to enhance the response amplitude. Especially, the functions of PD location and accurate measurement and even distributed monitoring are enhanced through various multiplexing combinations. To get an overall evaluation of these techniques, analysis, and comparison of different optical structures in PD detection refer to Table 5.1.

Table 5.1 Analysis and comparison of different optical structures in partial discharge detection.

Methods	Principle	Basic topology	Function[a]	Cost	Arrangement	Size	Reliability	Location	Sensitivity
FBG	Bragg gratings		• PD source location • Abnormal temperature rise detection	High	Medium	Small	High	Medium	High
Fabry–Perot interference	Interference		• Abnormal temperature rise detection	Medium	Easy	Small	High	Difficult	High
Dual-beam interference	All-fiber interference		• PD source location • Distribution detection	Medium	Difficult	Medium	High	Easy	Low
			• PD source location • Distribution detection	Medium	Difficult	Medium	Medium	Easy	High
			• PD source location • Distribution detection	High	Medium	Medium	Low	Medium	Medium

Multiplexing technique	Combination of various sensors FBG FPI Dual-beams interference	LS — CP — FBG1 FBG2 ......FBGn — *Multiplexing of the same structure* — Filter — DAQ — PD	• PD source location • Distribution detection Abnormal temperature rise detection	Low	Difficult	Large	High	Easy	High
		LS — DAQ — DF — CP1 — CP2 — FRM *Multiplexing of the different structure*	• PD source location • Distribution detection Abnormal temperature rise detection	Low	Difficult	Large	Medium	Easy	High
Distributed optical sensing technique	Optical Time Domain Reflection	LD — Circulator ① ② ③ — Brillouin scattering — Sensing optical fiber — Detection unit *Brillouin scattering*	• Long-distance detection • PD source location • Temperature distribution	High	Difficult	Small	Medium	Easy	High

Abbreviations in basic topology: LS: light source; P-D: photodiode; CP: coupler; SF: sensing fiber; RF: reference fiber; FRM: Faraday rotating mirror.

a) The function in common: measurement of PD-induced acoustic emission.

References

1 Gu, F., Chen, H., and Chao, M. (2018). Application of improved Hilbert-Huang transform to partial discharge signal analysis. *IEEE Transactions on Dielectrics and Electrical Insulation* 25 (2): 668–677.

2 Rodrigo, A., Llovera, P., Fuster, V., and Quijano, A. (2012). Study of partial discharge charge evaluation and the associated uncertainty by means of high frequency current transformers. *IEEE Transactions on Dielectrics and Electrical Insulation* 19 (2): 434–442.

3 Aizpurua, J., Catterson, V., Stewart, B. et al. (2018). Power transformer dissolved gas analysis through Bayesian networks and hypothesis testing. *IEEE Transactions on Dielectrics and Electrical Insulation* 25 (2): 494–506.

4 Stone, G.C. (2012). A perspective on online partial discharge monitoring for assessment of the condition of rotating machine stator winding insulation. *IEEE Electrical Insulation Magazine* 28 (5): 8–13.

5 Montanari, G. and Cavallini, A. (2013). Partial discharge diagnostics: from apparatus monitoring to smart grid assessment. *IEEE Electrical Insulation Magazine* 29 (3): 8–17.

6 Wu, M., Cao, H., Cao, J. et al. (2015). An overview of state-of-the-art partial discharge analysis techniques for condition monitoring. *IEEE Electrical Insulation Magazine* 31 (6): 22–35.

7 Wild, G. and Hinckley, S. (2008). Acousto-ultrasonic optical fiber sensors: overview and state-of-the-art. *IEEE Sensors Journal* 8 (7): 1184–1193.

8 Kanakambaran, S., Sarathi, R., and Srinivasan, B. (2018). Robust classification of partial discharges in transformer insulation based on acoustic emissions detected using Fiber Bragg gratings. *IEEE Sensors Journal* 18 (24): 10018–10027.

9 Ghorat, M., Gharehpetian, G., Latifi, H. et al. (2018). Partial discharge acoustic emission detector using mandrel-connected fiber Bragg grating sensor. *Optical Engineering* 57 (7): 074107.

10 Ghorat, M., Gharehpetian, G., Latifi, H. et al. (2019). High-resolution FBG-based fiber-optic sensor with temperature compensation for PD monitoring. *Sensors* 19 (23): 5285.

11 S. Jinachandran, "Design and development of a novel packaged fibre Bragg grating based acoustic emission monitoring system for crack detection in engineering applications," Thesis, University of Wollongong, 2020. https://ro.uow.edu.au/theses1/770/.

12 Rosenthal, A., Razansky, D., and Ntziachristos, V. (2011). High-sensitivity compact ultrasonic detector based on a pi-phase-shifted fiber Bragg grating. *Optics Letters* 36 (10): 1833–1835.

13 Zheng, Q., Ma, G., Jiang, J. et al. (2015). A comparative study on partial discharge ultrasonic detection using fiber Bragg grating sensor and piezoelectric transducer. In: *2015 IEEE Conference on Electrical Insulation and Dielectric Phenomena (CEIDP)*, 282–285. IEEE.

14 Kanakambaran, S., Sarathi, R., and Srinivasan, B. (2017). Identification and localization of partial discharge in transformer insulation adopting cross recurrence plot analysis of acoustic signals detected using fiber Bragg gratings. *IEEE Transactions on Dielectrics and Electrical Insulation* 24 (3): 1773–1780.

15 Ye, H., Qian, Y., Liu, Y. et al. (2014). Partial discharge detection technology based on the fibre Bragg grating. *Australian Journal of Electrical and Electronics Engineering* 11 (2): 217–225.

16 Haifeng, Y., Yong, Q., Yadong, L. et al. (2015). Partial discharge detection technology based on Fiber Bragg grating. *High Voltage Engineering* 41 (1): 6.

17 Perez, I., Cui, H., and Udd, E. (2001). Acoustic emission detection using fiber Bragg gratings. In: *SPIE's 8th Annual International Symposium on Smart Structures and Materials*, 4328. SPIE.

18 Ma, G., Zhou, H., Shi, C. et al. (2018). Distributed partial discharge detection in a power transformer based on phase-shifted FBG. *IEEE Sensors Journal* 18 (7): 2788–2795.

19 Shi, C., Ma, G., Mao, N. et al. (2017). Ultrasonic detection coherence of fiber Bragg grating for partial discharge in transformers. In: *2017 IEEE 19th International Conference on Dielectric Liquids (ICDL)*, 1–4. IEEE.

20 Wei, P., Han, X., Xia, D. et al. (2018). Novel Fiber-optic ring acoustic emission sensor. *Sensors* 18 (1): 215.

21 Islam, M., Ali, M., Lai, M. et al. (2014). Chronology of Fabry–Perot interferometer fiber-optic sensors and their applications: a review. *Sensors* 14 (4): 7451–7488.

22 Gao, C., Yu, L., Xu, Y. et al. (2019). Partial discharge localization inside transformer windings via fiber-optic acoustic sensor array. *IEEE Transactions on Power Delivery* 34 (4): 1251–1260.

23 Gao, C., Wang, W., Song, S. et al. (2018). Localization of partial discharge in transformer oil using Fabry-Pérot optical fiber sensor array. *IEEE Transactions on Dielectrics and Electrical Insulation* 25 (6): 2279–2286.

24 LeBlanc, M. (2000). Acoustic sensing using free and transducer-mounted fiber Bragg gratings. In: *Fourteenth International Conference on Optical Fiber Sensors*, vol. 4185, 41855E. International Society for Optics and Photonics.

25 Lima, S., Frazão, O., Araújo, F. et al. (2008). Fibre Fabry-Perot sensor for acoustic detection. In: *19th International Conference on Optical Fibre Sensors*, vol. 7004, 700441. International Society for Optics and Photonics.

26 Deng, J., Xiao, H., Huo, W. et al. (2001). Optical fiber sensor-based detection of partial discharges in power transformers. 33 (5): 305–311.

27 Cui, Q., Thakur, P., Rablau, C. et al. (2019). Miniature optical fiber pressure sensor with exfoliated graphene diaphragm. *IEEE Sensors Journal* 19 (14): 5621–5631.

28 Jung, I., Mallick, S., and Solgaard, O. (2009). A large-area high-reflectivity broadband monolithic single-crystal-silicon photonic crystal mirror MEMS scanner with low dependence on incident angle and polarization. *IEEE Journal of Selected Topics in Quantum Electronics* 15 (5): 1447–1454.

29 Luo, X., Tsai, D., Gu, M., and Hong, M. (2018). Subwavelength interference of light on structured surfaces. *Advances in Optics and Photonics* 10 (4): 757–842.

30 Ezbiri, A. and Tatam, R. (1996). Interrogation of low finesse optical fibre Fabry–Pérot interferometers using a four wavelength technique. *Measurement Science and Technology* 7 (2, 117).

31 Jia, P. and Wang, D. (2012). Self-calibrated non-contact fibre-optic Fabry–Perot interferometric vibration displacement sensor system using laser emission frequency modulated phase generated carrier demodulation scheme. *Measurement Science and Technology* 23 (11): 115201.

32 Tseng, S. and Chen, C. (1988). Optical fiber Fabry–Perot sensors. *Applied Optics* 27 (3): 547–551.

33 Frazão, O., Silva1, S., Viegas, J. et al. (2010). A hybrid Fabry–Perot/Michelson interferometer sensor using a dual asymmetric core microstructured fiber. *Measurement Science and Technology* 21 (2): 025205.

34 Farahi, F., Newson, T., Jones, J., and Jackson, D. (1988). Coherence multiplexing of remote fibre optic Fabry–Perot sensing system. *Optics Communications* 65 (5): 319–321.

35 Farahi, F. (1991). Fiber-optic interferometric point thermometer. *Fiber & Integrated Optics* 10 (2): 205–212.

36 Si, W., Fu, C., Li, D. et al. (2018). Directional sensitivity of a mems-based fiber-optic extrinsic fabry–perot ultrasonic sensor for partial discharge detection. *Sensors* 18: 6–1975.

37 Chen, Q., Zhang, W., and Zhao, H. (2019). Response bandwidth design of Fabry–Perot sensors for partial discharge detection based on frequency analysis. *Journal of Sensors* 2019: 1026934.

38 Hayber, S.E., Tabaru, T.E., and Saracoglu, O. (2019). A novel approach based on simulation of tunable MEMS diaphragm for extrinsic Fabry–Perot sensors. *Optics Communications* 430: 14–23.

39 Yan, L., Gui, Z., Wang, G. et al. (2017). A micro bubble structure based Fabry–Perot optical fiber strain sensor with high sensitivity and low-cost characteristics. *Sensors* 17 (3): 555.

40 Zhang, Y., Yuan, L., Lan, X. et al. (2013). High-temperature fiber-optic Fabry–Perot interferometric pressure sensor fabricated by femtosecond laser: erratum. *Optics Letters* 38 (22): 4609–4612.

41 Beard, P., Perennes, F., and Mills, T. (1999). Transduction mechanisms of the Fabry–Perot polymer film sensing concept for wideband ultrasound detection. *IEEE Transactions on Ultrasonics, Ferroelectrics, and Frequency Control* 46 (6): 1575–1582.

42 Gong, Z., Chen, K., Zhou, X. et al. (2017). High-sensitivity Fabry-Perot interferometric acoustic sensor for low-frequency acoustic pressure detections. *Journal of Lightwave Technology* 35 (24): 5276–5279.

43 Wang, X., Li, B., Xiao, Z. et al. (2004). An ultra-sensitive optical MEMS sensor for partial discharge detection. *Journal of Micromechanics and Microengineering* 15 (3): 521.

44 Wang, K., Tong, X., and Zhu, X. (2014). Transformer partial discharge monitoring based on optical fiber sensing. *Photonic Sensors* 4 (2): 137–141.

45 Zhang, W., Chen, Q., Zhang, L., and Hong, Z. (2018). Fiber Optic Fabry-Perot Sensor with Stabilization Technology for Acoustic Emission Detection of Partial Discharge. In: *2018 IEEE International Conference on High Voltage Engineering and Application (ICHVE)*, 1–4. IEEE.

46 Wang, Y., Li, X., Gao, Y. et al. (2018). Partial discharge ultrasound detection using the Sagnac interferometer System. *Sensors* 18 (5): 1425.

47 Liu, H. (2016). Acoustic partial discharge localization methodology in power transformers employing the quantum genetic algorithm. *Applied Acoustics* 102: 71, 78.

48 Ashraf, S., Brian, G., Zhou, C., and Jahabar, J. (2006). Modelling of acoustic signals from Partial discharge activity. In: *2006 IEEE GCC Conference (GCC)*, 1–5. IEEE.

49 Wotzka, D. (2012). Mathematical description of acoustic emission signals generated by partial discharges. In: *2012 International Conference on High Voltage Engineering and Application*, 617–620. IEEE.

50 Zhou, H., Ma, G., Wang, Y. et al. (2019). Optical sensing in condition monitoring of gas insulated apparatus: a review. *High Voltage* 4 (4): 259–270.

51 Xu, Y., Lu, P., Qin, Z. et al. (2013). Vibration sensing using a tapered bend-insensitive fiber based Mach-Zehnder interferometer. *Optics Express* 21 (3): 3031–3042.

52 Lee, B., Kim, Y., Park, K. et al. (2012). Interferometric fiber optic sensors. *Sensors* 12 (3): 2467–2486.

53 Lu, H., Wang, X., Zhang, S. et al. (2018). A fiber-optic sensor based on no-core fiber and Faraday rotator mirror structure. *Optics & Laser Technology* 101: 507, 514.

54 Vali, V. and Shorthill, R. (1976). Fiber ring interferometer. *Applied Optics* 15 (5): 1099–1100.

55 Cho, L.-H., Wu, C., Lu, C., and Tam, H. (2013). A highly sensitive and low-cost Sagnac loop based pressure sensor. *IEEE Sensors Journal* 13 (8): 3073–3078.

56 Jiang, J., Wang, K., Wu, X. et al. (2020). Characteristics of the propagation of partial discharge ultrasonic signals on a transformer wall based on Sagnac interference. *Plasma Science and Technology* 22 (2): 024002.

57 Liu, Q., Li, S.-G., and Wang, X. (2017). Sensing characteristics of a MF-filled photonic crystal fiber Sagnac interferometer for magnetic field detecting. *Sensors and Actuators B: Chemical* 242: 949–955.

58 Macià-Sanahuja, C., Lamela, H., and García-Souto, J. (2007). Fiber optic interferometric sensor for acoustic detection of partial discharges. *Journal of Optical Technology* 74 (2): 122–126.

59 Zargari, A. and Blackbum, T. (2001). A non-invasive optical fibre sensor for detection of partial discharges in SF/sub 6/-GIS systems. In: *Proceedings of 2001 International Symposium on Electrical Insulating Materials (ISEIM 2001). 2001 Asian Conference on Electrical Insulating Diagnosis (ACEID 2001). 33rd Symposium on Electrical and Ele*, 359–362. IEEE.

60 Zhang, T., Pang, F., Liu, H. et al. (2016). A fiber-optic sensor for acoustic emission detection in a high voltage cable system. *Sensors* 16 (12): 2026.

61 Zhang, A. and Zhang, S. (2016). High stability fiber-optics sensors with an improved PGC demodulation algorithm. *IEEE Sensors Journal* 16 (21): 7681–7684.

62 Kishore, P.N., Shankar, M.S., Kishore, P. et al. (2013). Analyzing characteristics of Sagnac loop interferometric stress sensor. In: *2013 International Conference on Microwave and Photonics (ICMAP)*, 1–3. IEEE.

63 Qian, S., Chen, H., Xu, Y. et al. (2016). Acoustic fiber optic sensors for partial discharge monitoring. In: *2016 IEEE Electrical Insulation Conference (EIC)*, 109–112. IEEE.

64 Qian, S., Chen, H., Xu, Y., and Su, L. (2018). High sensitivity detection of partial discharge acoustic emission within power transformer by sagnac fiber optic sensor. *IEEE Transactions on Dielectrics and Electrical Insulation* 25 (6): 2313–2320.

65 Joe, H.-E., Yun, H., Jo, S.-H. et al. (2018). A review on optical fiber sensors for environmental monitoring. *International Journal of Precision Engineering and Manufacturing – Green Technology* 5 (1): 173–191.

66 Koelling, M., Meinl, T., Malik, Y. et al. (2018). Acoustic partial discharge measurements on medium voltage cable connectors using fiber optic sensors. In: *VDE High Voltage Technology 2018; ETG-Symposium*, 1–5. VDE.

67 Wang, T., Luo, C., and Zheng, S. (2001). A fiber-optic current sensor based on a differentiating Sagnac interferometer. *IEEE Transactions on Instrumentation and Measurement* 50 (3): 705–708.

68 Jang, T., Lee, S., and Kim, Y. (2004). Surface-bonded fiber optic Sagnac sensors for ultrasound detection. *Ultrasonics* 42 (1–9): 837–841.

69 Muanenda, Y., Faralli, S., Oton, C.J., and Pasquale, F. (2018). Dynamic phase extraction in a modulated double-pulse φ-OTDR sensor using a stable homodyne demodulation in direct detection. *Optics Express* 26 (2): 687–701.

70 Dandridge, A., Tveten, A., and Giallorenzi, T. (1982). Homodyne demodulation scheme for fiber optic sensors using phase generated carrier. *IEEE Transactions on Microwave Theory and Techniques* 30 (10): 1635–1641.

71 He, X., Xie, S., Liu, F. et al. (2017). Multi-event waveform-retrieved distributed optical fiber acoustic sensor using dual-pulse heterodyne phase-sensitive OTDR. *Optics Letters* 42 (3): 442–445.

72 Fang, G., Xu, T., and Li, F. (2015). Heterodyne interrogation system for TDM interferometric fiber optic sensors array. *Optics Communications* 341: 74–78.

73 Zhang, W., Gao, W., Huang, L. et al. (2015). Optical heterodyne micro-vibration measurement based on all-fiber acousto-optic frequency shifter. *Optics Express* 23 (13): 17576–17583.

74 Fang, W., Jia, Q., Zhen, S. et al. (2015). Low coherence fiber differentiating interferometer and its passive demodulation schemes. *Optical Fiber Technology* 21: 34–39.

75 Yang, X., Chen, Z., Ng, J.H. et al. (2012). A PGC demodulation based on differential-cross-multiplying (DCM) and arctangent (ATAN) algorithm with low harmonic distortion and high stability. In: *OFS2012 22nd International Conference on Optical Fiber Sensors*, vol. 8421, 84215J. International Society for Optics and Photonics.

76 Nikitenko, A., Plotnikov, M., Volkov, A. et al. (2018). PGC-Atan demodulation scheme with the carrier phase delay compensation for fiber-optic interferometric sensors. *IEEE Sensors Journal* 18 (5): 1985–1992.

77 Zhang, S., Chen, B., Yan, L., and Xu, Z. (2018). Real-time normalization and non-linearity evaluation methods of the PGC-arctan demodulation in an EOM-based sinusoidal phase modulating interferometer. *Optics Express* 26 (2): 605–616.

78 Volkov, A., Plotnikov, M., Mekhrengin, M. et al. (2017). Phase modulation depth evaluation and correction technique for the PGC demodulation scheme in fiber-optic interferometric sensors. *IEEE Sensors Journal* 17 (13): 4143–4150.

79 Wang, G.-q., Xu, T.-W., and Li, F. (2012). PGC demodulation technique with high stability and low harmonic distortion. *IEEE Photonics Technology Letters* 24 (23): 2093–2096.

80 Zhang, W., Xia, H., Pan, C. et al. (2014). Differential-self-multiplying-integrate phase generated carrier method for fiber optic sensors. In: *International Symposium* *on Photonics and Optoelectronics 2014*, vol. 9233, 92331U. International Society for Optics and Photonics.

81 Zhang, A. and Li, D. (2018). Interferometric sensor with a PGC-AD-DSM demodulation algorithm insensitive to phase modulation depth and light intensity disturbance. *Applied Optics* 57 (27): 7950–7955.

82 Wang, C., Wang, C., Shang, Y. et al. (2015). Distributed acoustic mapping based on interferometry of phase optical time-domain reflectometry. *Optics Communications* 346: 172–177.

83 Wang, C., Shang, Y., Liu, X. et al. (2014). Distributed acoustic mapping based on self-interferometry of phase-OTDR. In: *2014 Asia Communications and Photonics Conference (ACP)*, 1–3. IEEE.

84 Liao, H., Liao, H., Lu, P. et al. (2017). Phase demodulation of short-cavity Fabry–Perot interferometric acoustic sensors with two wavelengths. *IEEE Photonics Journal* 9 (2): 1–9.

85 Lamela, H., Garcia-Souto, J.A., and Macia-Sanahuja, C. (2005). Interferometric optical fiber sensors for measurements within oil-filled power transformers. In: *Optical Fibers: Applications*, vol. 5952, 595209. International Society for Optics and Photonics.

86 Russell, S., Brady, K., and Dakin, J.P. (2001). Real-time location of multiple time-varying strain disturbances, acting over a 40-km fiber section, using a novel dual-Sagnac interferometer. *Journal of Lightwave Technology* 19 (2): 205–213.

87 Sarkar, B., Mishra, D., Koley, C. et al. (2016). Intensity-modulated fiber Bragg grating sensor for detection of partial discharges inside high-voltage apparatus. *IEEE Sensors Journal* 16 (22): 7950–7957.

88 Comanici, M., Zhang, L., Chen, L. et al. (2012). All-Fiber DBR-Based Sensor Interrogation System for Measuring Acoustic Waves. *Journal of Sensors* 2012.

89 Wang, J., Ai, F., Sun, Q. et al. (2018). Diaphragm-based optical fiber sensor array for multipoint acoustic detection. *Optics Express* 26 (19): 25293–25304.

90 Zhu, P., Wen, H., Che, Q. et al. (2019). Disturbed PD Detection System Based on Improved φ-OTDR Assisted by wFBG Array. In: *2019 18th International Conference on Optical Communications and Networks (ICOCN)*, 1–3. IEEE.

91 D. Srivastava, I. Bhutani, and B. Das, "Experimental realization of a fiber based Michelson interferometer for FBG sensor interrogation." *International Symposium on Photonics (OSI - ISO)*, India, 2018.

92 Lima, S., Frazao, O., Farias, R. et al. (2010). Mandrel-based fiber-optic sensors for acoustic detection of partial discharges—A proof of concept. *IEEE Transactions on Power Delivery* 25 (4): 2526–2534.

93 Lima, S., Frazao, O., Farias, R. et al. (2010). Fibre laser sensor based on a phase-shifted chirped grating for acoustic sensing of partial discharges in power transformers. In: *Fourth European Workshop on Optical Fibre Sensors*, vol. 7653, 765335. International Society for Optics and Photonics.

94 Kim, H., Sampath, U., and Song, M. (2015). Multi-stress monitoring system with fiber-optic mandrels and fiber Bragg grating sensors in a Sagnac loop. *Sensors* 15 (8): 18579–18586.

95 Lee, J. (2012). Sound detection monitoring in the transformer oil using fiber optic sensor. In: *Fiber Optic Sensors and Applications IX*, vol. 8370, 83700W. International Society for Optics and Photonics.

96 Li, M. and Zhao, H. (2007). All-fiber system based on Fabry-Perot sensor for partial discharges in transformer oil. In: *Fundamental Problems of Optoelectronics and Microelectronics III*, vol. 6595, 65952I. International Society for Optics and Photonics.

97 Lima, S.E., Frazao, O., Farias, R. et al. (2009). Fiber Fabry–Perot sensors for acoustic detection of partial discharges in transformers. In: *2009 SBMO/IEEE MTT-S International Microwave and Optoelectronics Conference (IMOC)*, 307–311. IEEE.

98 Yin, Z., Zhang, R., Tong, J., and Chen, X. (2013). An all-fiber partial discharge monitoring system based on both intrinsic fiber optic interferometry sensor and fluorescent fiber. In: *2013 International Conference on Optical Instruments and Technology: Optical Sensors and Applications*, vol. 9044, 904414. International Society for Optics and Photonics.

99 Delepine-Lesoille, S., Merliott, E., Boulay, C. et al. (2006). Quasi-distributed optical fibre extensometers for continuous embedding into concrete: design and realization. *Smart Materials and Structures* 15 (4): 931.

100 Mahanta, D. and Laskar, S. (2016). Transformer condition monitoring using fiber optic sensors: a review. *ADBU Journal of Engineering Technology* 4.

101 Rohwetter, P., Eisermann, R., and Krebber, K. (2015). Distributed acoustic sensing: Towards partial discharge monitoring. In: *24th International Conference on Optical Fibre Sensors*, vol. 9634, 96341C. International Society for Optics and Photonics.

102 Zhu, P., Wen, H., Che, Q., and Li, X. (2020). Disturbed partial discharge detection system based on an improved Φ-OTDR assisted by a wFBG array. *Applied Optics* 59 (14): 4367–4370.

103 Wan, X., Du, T., Zhang, Z. et al. (2013). Positioning approach based on Mach-Zehnder fiber sensors and a DSP processor. In: *2013 International Conference on Optical Instruments and Technology: Optical Sensors and Applications*, vol. 9044, 90440T. International Society for Optics and Photonics.

104 Zhou, H.-y., Ma, G.-m., Zhang, M. et al. (2019). A high sensitivity optical fiber Interferometer sensor for acoustic emission detection of partial discharge in power transformer. *IEEE Sensors Journal*.

105 Luo, J., Hao, Y., Ye, Q. et al. (2013). Development of optical fiber sensors based on Brillouin scattering and FBG for on-line monitoring in overhead transmission lines. *Journal of Lightwave Technology* 31 (10): 1559–1565.

106 Masoudi, A., Belal, M., and Newson, T. (2013). Distributed dynamic large strain optical fiber sensor based on the detection of spontaneous Brillouin scattering. *Optics Letters* 38 (17): 3312–3315.

107 Liu, R., Babanajad, S., Taylor, T., and Ansari, F. (2015). Experimental study on structural defect detection by monitoring distributed dynamic strain. *Smart Materials and Structures* 24 (11): 115038.

108 Palmieri, L. and Schenato, L. (2013). Distributed optical fiber sensing based on Rayleigh scattering. *The Open Optics Journal* 7 (1).

109 Bado, M., Casas, J., and Barrias, A. (2018). Performance of Rayleigh-based distributed optical fiber sensors bonded to reinforcing bars in bending. *Sensors* 18 (9): 3125.

110 Ding, Z., Yang, D., Du, Y. et al. (2016). Distributed strain and temperature discrimination using two types of fiber in OFDR. *IEEE Photonics Journal* 8 (5): 6804608.

6

Other Parameters with Optical Methods

Apart from the typical parameters mentioned previously, some mechanical detection and electrical measurements can be achieved through optical techniques for the harsh conditions in a power transformer application. Winding deformation and vibration detection are of great interest to be monitored in the field, closely related to the mechanical status of the transformer. Since optical methods are also well recognized in mechanical detection in various structural health monitoring (SHM), they offer wonderful solutions to determine mechanical defects in the transformer. Voltage, current, and the electric field are very important but are basic electric parameters for the design and monitoring of power transformers, which also can be available through optical techniques.

In the end, optical methods applied in power transformers have been critically concluded and discussed as a whole. Technical barriers and prospects of optical monitoring methods for power transformers are presented as well.

6.1 Winding Deformation and Vibration Detection in Optical Techniques

6.1.1 Winding Deformation Detection

6.1.1.1 Winding Deformation in Power Transformer

Owing to the fact that the overall short-circuit resistance of the transformer is insufficient, 40% of them are damaged when subjected to a short-circuit impact. A huge electromagnetic force (EMF) in short periods originating from the intense currents flowing in the transformer windings, induced by internal faults in the transformers or by external faults in the network to which the transformer is connected, are sufficient to mechanically deform or damage the windings. Therefore, the transformer windings experience mechanical damages due to this impact, such as axial displacement, bending, or tilting of windings due to axial forces and winding deformation due to radial forces. If the transformer continues to operate without detecting winding deformation, its cumulative effect will gradually develop, increasing damage to the life and reliability of the transformer It may also cause it to withdraw from the operation, giving rise to a large-scale power outage and other events. Therefore, an accurate estimation of the operating state of the transformer, especially the health of the windings and the

Optical Sensing in Power Transformers, First Edition. Jun Jiang and Guoming Ma.
© 2021 John Wiley & Sons Ltd. Published 2021 by John Wiley & Sons Ltd.

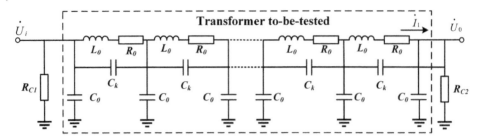

Figure 6.1 Equivalent circuit of the sweep frequency impedance method at a high frequency band. Source: Adapted from Yong et al. [3].

timely diagnosis and maintenance of possible faults, can effectively reduce the defect probability in the transformer.

Frequency response analysis (FRA) is proven to be a powerful and widely accepted tool to detect winding deformation within power transformers [1, 2]. The working principle of FRA is simple. Specifically, once the load voltage frequency is higher than 1 kHz, the excitation effect on the transformer core disappears, and the winding can be regarded as a linear circuit composed of a series of distributed parameters such as inductance, capacitance, and resistance (the resistance is very small and can be ignored), as shown in Figure 6.1.

A functional relationship through the transfer function characteristic curve is established between the input excitation and the output response, and a set of amplitude-frequency and phase-frequency characteristics of the tested winding can be obtained. If the distribution parameters of the transformer in this circuit change, the sweep frequency impedance value will inevitably change. Therefore, the sweep frequency impedance curve is similar to the frequency response curve and describe the interturn short circuit of the transformer winding deformation.

However, there still exist problems regarding the application of FRA. FRA is a comparative method in which the measured FRA signature should be compared with its fingerprint. The resolution is limited and results are hard to interpret since there is no reliable standard code for FRA signature classification and quantification. Moreover, FRA cannot be used as an online tool as it is impossible to detect dynamic electromagnetic and mechanical forces interaction.

Since an optical fiber is passive and immune against electromagnetic interference, it is possible to measure the displacements or strains in power transformer windings directly. Thus, the optical solution stands out as flexible and particularly suitable for the early warning or alert of mechanical defects like winding looseness, displacement, deformation, or vibration.

6.1.1.2 Winding Deformation Detection with an Optical Technique

Since the dynamic force on the windings in a power transformer reaches peaks as high as 100 kN or an equivalent pressure of 14 MPa on the condition of a short-circuit current, then the winding deformation can be the stress-based effect or pressure [4–7].

Strain is one of the key sensitive parameters for a typical fiber Bragg grating (FBG) structure, and it is easy to measure the pressure in that sense. However, the design and installation of the FBG pressure probe are the critical points.

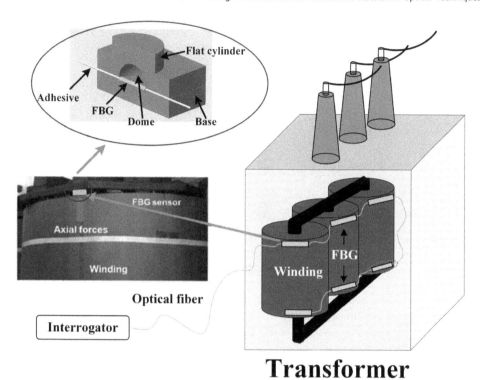

Figure 6.2 Deformation detection in transformer windings with optical sensors. Source: Adapted with permission from Melo et al. [7].

A typical sensor conception is depicted in Figure 6.2, including the base (substrate), dome, adhesive material, and flat cylindrical unit. Regarding the base, the shape is designated to the placement of the winding, and its material is anticipated to be stable at high temperatures and sensitive to the pressure in an appropriate margin. In order to obtain a better sensitivity when axial forces are produced in the windings, some modifications are inserted in the shim shape like a rectangular block with an empty dome. Also, a flat cylindrical shaped element is fixed on the top, which is beneficial to the axial forces toward the dome. A high temperature-resistant epoxy adhesive is recommended as well. The designed FBG sensors can then be inserted in the top or in the base of the transformer windings to monitor the vibration signal and variations inside the transformer.

Nowadays, the distributed optical sensing is also available to transmit wire strain efficiently for distribution curves of the transformer winding and even the location of the mechanical winding defect [8, 9]. Based on the scattering effect, like those of Raman or Brillouin or Rayleigh, the related information has been provided and discussed in distributed temperature sensing in Chapter 2.

As to distributed sensing, the optical fiber arrangement is one of the key techniques. Regarding fiber flange and penetration, two aspects are weighed as on-site installation and actual detection. It is supposed to select the installation of the fiber penetrator and flange on the transformer tank wall near the low voltage bushing on the tank wall. As to the wiring of the optical fiber inside the transformer, both operation reliability and compatibility with the insulation structure need to be determined. The installation of

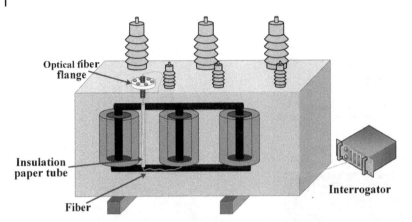

Optical fiber flange

Insulation paper tube

Fiber

Interrogator

Figure 6.3 Optical fiber arrangement in the transformer. Source: Adapted from Yuang [10].

the optical fiber penetrates the wall and the transmission fiber is compatible with the electric field distribution inside the transformer. It is necessary to avoid excessive tension, knotting, torque, and winding of the fiber in order to decrease transmission loss. The internal optical fibers are expected to converge in a shared insulation tube from the optical fiber penetration, as shown in Figure 6.3.

Since the electric field intensity of the grounding terminal is bound to be small, it is recommended that the distributed fiber should be arranged on the low voltage winding side.

However, the scheme is good for installation in a newly designed transformer prior to being applied onsite to provide high spatial resolution and a wide measurement range.

6.1.2 Vibration Detection

6.1.2.1 Vibration in Power Transformers

The practical operation experiences and accidental disintegration cases of worldwide transformers indicate that the mechanical defect of the winding is one of the important causes of the failure of the transformer. Therefore, the research on the detection and diagnosis of the mechanical state of the winding becomes a hot spot in the field. Since the 1980s, researchers around the world have successively proposed FRA, a low voltage pulse method, the vibration signal analysis method, and the frequency sweep impedance method, etc., to detect the mechanical state of the windings. The vibration phenomenon and resultant signal analysis, in particular, have received extensive attention because of the contactless electrical connection with the transformer and easy implementation of online or live detection. Therefore, the transformer tank vibration technique is considered to be an effective and promising tool to diagnose power transformers. Both the internal faults or unusual operating machine conditions can be measured by comparing the fingerprint vibration data of the transformer tank, initially acquired, to the last measurement results [11, 12].

The vibrations on the surface of the transformer oil tank mainly come from the magnetostrictive effect of the transformer core and the forced vibration caused by the periodic force by the alternating leakage magnetic field of the winding. When the mechanical state of the transformer winding changes, the vibration status also becomes different,

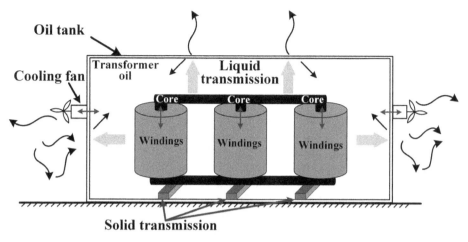

Figure 6.4 Schematic diagram of transformer vibration transmission and noise radiation. Source: Adapted from Shengchang et al. [14].

and eventually propagates to the surface of the oil tank, thus providing a basis for the mechanical state of the winding by the vibration technique [13].

With regards to the source and distribution of vibration in a typical oil-immersed power transformer, the vibration of the core, winding, and oil pump is transmitted to the surface of the tank through the transformer oil and fasteners, and then radiates noise outside. The cooling fan also affects the vibration and noise of the transformer through fasteners and air pulsations, as depicted in Figure 6.4.

Under the no-load state, the winding vibration can be ignored because of the small excitation current that flows in the winding. Therefore, the vibration of the surface of the oil tank during no-load is just the vibration of the iron core, which can also be called the no-load vibration. During the steady-state short-circuit experiment (i.e. load experiment), the voltage is very low, the magnetic flux in the core is very small, and its vibration can be ignored. Therefore, the vibration of the transformer at this time is mainly superposition of the winding, the magnetic shield, and the vibration of the tank wall. It can then be named the load vibration status. In an abnormal state, the winding deformation or unexpected winding movement (up to a few nanometers), induced by transportation damage or large current surges like lightning strikes/short circuit faults, and the mechanical defects can be reflected through the vibration detection. It is recognized that the vibration signatures inside the oil transformer have a wider band frequency, from 20 to 1000 Hz, where the frequency gradually attenuates to zero after more than 1000 Hz [15]. Accompanied by vibration, sound is also proved to be an effective parameter to weigh some mechanical defects, with a higher frequency range up to 20 kHz.

An FRA-like technique helps to diagnose the structural defects to some extent after a comparison with the offline referenced response spectrum, but the online measurement is more appropriate to excite the mechanical defects in practical applications. Fiber optical sensing also provides an alternative solution to the online detection of vibration for power transformers.

Conventional vibration techniques, on the basis of capacitive or piezoelectric transducers, are limited in application to the presence of high voltages or high magnetic fields due to the problem of electrical isolation. Also, excessively high or low atmospheres, e.g. SF_6 or a vacuum in electrical circuit-breakers. Fiber-optic-based instrumentation is therefore an attractive alternative to vibration measurements in the vicinity of an electrical substation including power transformers.

6.1.2.2 Vibration Detection with Optical Techniques

Interferometric-Based Vibration Measurement The vibration measurement is quite similar to the description of optical partial discharge on the basis of an acoustic emission effect described in Chapter 5. However, different from that of a partial discharge, the frequency spectrum of the vibrations located in a low frequency range is mainly less than 2 kHz, and the features of the harmonics at power frequency (50/60 Hz) are clear. Moreover, the amplitude of the vibration signal is significantly higher than that of PD. In other words, the requirements of vibration detection are much easier than that of PD. Thus, most optical PD techniques can be transferred to vibration measurements, such as interferometric-based techniques and FBG-based methods [16].

As to interferometric techniques, they are good at interrogating frequency-dependent signals, both Fabry Perot interferometry [17] and fiber interferometric are feasible to measure vibrations of power transformers, and the latter is the mainstream.

Typically, a fiber-optic multi-channel Mach-Zehnder interferometric sensor is proposed to detect the characteristic frequencies for the vibration (100 Hz or harmonics of 100 Hz), as depicted in Figure 6.5. There are a reference coil and sensing arms to make the comparison. The paths of the sensing fibers are common for addressing the monitored region with reduced disturbance. The interference output is proportional to the cosine of the optical phase and the optical phase shift with time is just in response to the vibrations. Since the quasi-static initial conditions relate to the initial optical phase and changes as the low frequency drifts, it is necessary to separate them. The interference signal is then forced to be multi-period in response to the vibrations [18]. The sensing probe design and installation is important to the optical measurement system.

The high-sensitivity intrinsic probe of the optical fiber can be installed at the magnetic core of power transformers. Prior to installation, the calibration with factors such as dynamic strain, frequency response of vibrations, and the disturbance of temperature

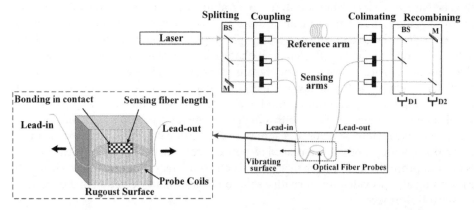

Figure 6.5 Optoelectronic setup of the fiber-optic laser interferometer. Source: Adapted from Garcia-Souto et al. [18].

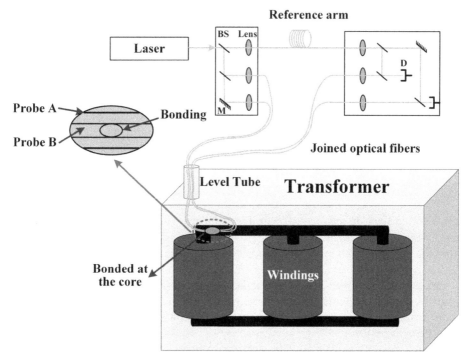

Figure 6.6 Optoelectronic setup for interferometric measurements inside the power transformer. Source: Adapted from Rivera et al. [19].

variations are confirmed in the laboratory. The dynamic strain can then be dynamically modulated and detected with regard to the magnetostriction of the core through changes in the optical path length (Figure 6.6).

The probe can be completely bonded to monitor dynamic strain. Optical fibers are able to access the sensing region and satisfy these environmental conditions of high electric magnetic fields and a wide range of high temperatures, especially in the load. Moreover, the discrimination between different sources of vibrations (core and windings) is also available in contact with the to-be-monitored structure or section.

Similarly, the Fabry–Perot interferometer (FPI) also provides a good alternative solution for vibration sensing [17].

FBG-Based Vibration Measurement FBG is also a good candidate for the demodulation of dynamic strain, especially in a relatively low frequency compared to partial discharge. Regarding the commercial high speed demodulation interrogator, the frequency can cover the main characteristic domain of vibration in a power transformer.

The sensing mechanism and probe design are very critical when performing a vibration measurement. For example, a mechanical structure with the cantilever beam and the hinged unit is designed in Figure 6.7. Inspired by the acceleration sensor principle, the mass unit is connected to the wall though the metal hinge and there is also an elongated beam structure with one end of the fiber attached to the extension beam. The external microvibration motivates the probe and the mass acts as an inertial module, driving the mass to rotate around the hinge. Furthermore, the vibration can be detected

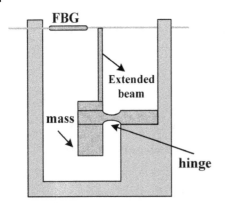

Figure 6.7 Typical structure of an FBG vibration sensor. Source: Adapted from Min et al. [15].

by the FBG unit and its change of central wavelength change, since the axial strain is forced by extension of the beam.

Since the fiber grating vibration sensor has a small volume, stable physical properties, and high insulation, it can also be closely combined with the power transformer. With appropriate packaging and insulating materials, the vibration sensor can fit between the narrow gaps of the windings inside the transformer and online detection of windings distortion. The vibration measurement, using a thin fiber optics sensor in power transformers, is available [16].

Fortunately, an optical fiber provides another immersed solution for the power transformer. As to the FBG-based vibration measurement, since FBG is mainly modulated by wavelength, the wavelength measurement resolution and detection speed are the critical factors for sensitivity. To improve the performance of the FBG-based vibration sensor, a subpicometer wavelength resolution measurement prototype has been developed with a compact volume and low cost [20]. Optical elements, with laterally varying transmission properties in conjunction with photo detectors and optical switches, are combined in the readout to obtain versatile parameters. Vibration and temperature are monitored using a multiplexed configuration of FBGs located at critical positions inside the power transformer, as shown in Figure 6.8. Real time on-line monitoring is achieved to monitor the vibration signal of the transformer, not only for the winding deformation fault but also for sensitivity to looseness of the structural parts such as the iron core and the tap changer.

Since the installation on the surface of anoil tank wall is the most convenient and easiest way to get access to the vibration information for the in-service transformer, the installation of the vibration sensors needs to be considered.

Installation Points of the Vibration Sensors   The placement of the sensors in a real transformer is proposed to occur in the outer winding and in the outer structure, due to feasibility of installation at these points. In the practical application of online vibration for the power transformer, fault diagnosis of the transformer winding is generally achieved by testing the vibration signal of the tank surface and analyzing its vibration characteristics. Therefore, it is crucial to obtain the vibration figures on the tank surface.

Single or a few measurement points are prone to large measurement errors when collecting the vibration signals. As it is difficult to fully and accurately reflect the state of the transformer winding, multiple measurement points are required. Since the power

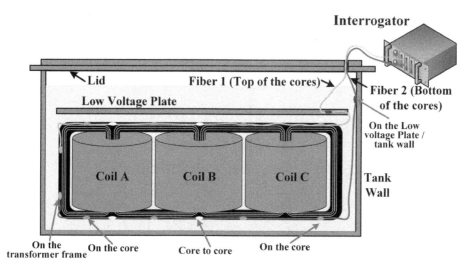

Figure 6.8 FBG installation for vibration monitoring in a transformer. Source: Adapted from Raghavan et al. [20].

transformer has a large volume, the vibrations of the internal windings are different at the time of the operation. Due to the inconsistency of the oil tank structure and the complexity of the winding vibration propagation process, the vibrations on the surface of the oil tank cannot effectively reflect the condition of the transformer windings. In addition, the vibration characteristics of the cabinet are often most related to the nearest vibration source. Installing vibration sensors at several key parts on the surface of the oil tank is a great necessity. By comparing and analyzing the vibration of different detection points, it is also possible to determine the location of the fault. Therefore, reasonable selection of the vibration measurement point is very important to evaluate the online status monitoring and fault diagnosis.

When conducting the transformer vibration test, the vibration detection point is recommended to be as close as possible to the transformer tank wall. The detection point corresponds to the upper and lower points of the winding. For larger transformers, the three points of upper, middle, and lower can be used. A typical vibration detection point arrangement of a three-phase transformer can be carried out as shown in Figure 6.9. This arrangement of measurement points is also applicable to the vibration test of a single-phase transformer.

6.1.3 Merits and Drawbacks

Winding deformation and vibration measurement are closely related. The former is the typical mechanical defect and the latter is a good indicator of mechanical faults like winding looseness, winding deformation, and insulation deterioration. This means that vibration is an effective and powerful approach to winding deformation. Various optical fiber techniques can be used to detect the mechanical defects (directly and indirectly) with several advantages.

Evidentially, optical sensors have a high sensitivity and flexible installation. Winding movements are hard to be detected as windings are always encased within the oil tank. Compared to a conventional FRA approach or visual inspection, fiber probes are

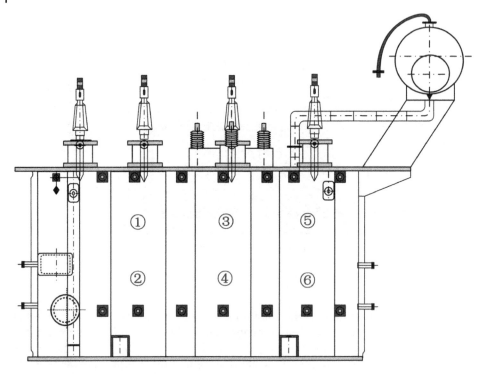

Figure 6.9 Typical vibration detection point arrangement of a three-phase transformer.

available to inside windings or other components in order to gain a higher sensitivity with more effective information. On the other hand, the demodulation at the vibration frequency is relatively easy with low cost. Since the main frequency of vibration is low, the interrogating scheme is simple to perform.

Every coin has two sides. The collection of vibration signals is inevitably affected by environmental noise, vibration interferences, etc., due to the complexity of the transformer working environment and its composition. Filtering and denoising of the original vibration signal is encouraged to be performed in order to get better results. In addition, the optical fiber is too fragile to directly detect the mechanical defects, like deformation, displacement, and vibration. Suitable packaging and reliable topology must be weighed from this perspective.

6.2 Voltage and Current Measurement with Optical Techniques

Voltage and current are the basic electric parameters for high voltage apparatus. Although voltage transducers (VTs) and current transducers (CTs) are designed and configured in typical power grid or substation, optical voltage transducers (OVTs) and optical current transducers (OCTs) have been put forward and potentially integrated on to power transformers due to their advantages of small volume and a wide dynamic measurement range from a future perspective. With this consideration, optical voltage and current measurements are involved in this chapter.

In theory, a light beam is reflected and transmitted in the medium surrounded by an electromagnetic field due to the influence of the electromagnetic field on the movement of electrons in the medium. Then the polarization characteristics of the original isotropic medium are changed, so that the polarization state of the light varies, generating a magneto-optical and electro-optic effect. The results of the interaction between the medium and the electromagnetic fields can be used to evaluate the measurement of the current and voltage using optical techniques.

6.2.1 Current Measurement with Optical Technique

6.2.1.1 Principle of Optical Current Transducer

Optical current transformers refer to those devices that comply with Ampere's law, through Faraday magneto-optic effects, to detect current by measuring the cumulative effect (integration) of the magnetic field in the optical loop around the measured current [21, 22].

When linearly polarized light passes through the medium under the magnetic field generated by the current, its polarization plane rotates. The deflection angle is proportional to the product of the magnetic induction intensity and the length of the optical path through the medium [23]:

$$\varphi = \mu V \int_L H(l) \cdot dl \tag{6.1}$$

where φ is the Faraday rotation angle; μ and V are the permeability and Verdet constant of the medium, respectively; H is the strength of the magnetic field; and L is the optical path length of polarized light passing through the medium. As the magnetic field intensity is generated by I, according to the Ampere loop law, Eq. (6.1) can be expressed as (see Figure 6.10):

$$\varphi = \mu V N I \tag{6.2}$$

Since it is currently impossible to achieve a high-precision measurement of the angle of the polarization plane, the change in the angle of the polarization plane is usually converted into a change in light intensity. Additionally, the polarization measurement and

Figure 6.10 Illustration of the Faraday magneto-optic effect.

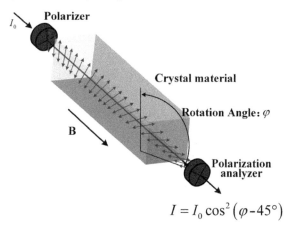

$$I = I_0 \cos^2\left(\varphi - 45°\right)$$

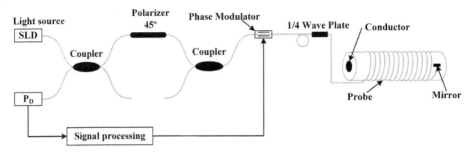

Figure 6.11 Structure and topology of an all-fiber type of optical current transformer. Source: Adapted from Silva et al. [24].

interference detection methods are used to achieve the current measurement. The Faraday magneto-optic effect has a good linear measurement, capable of changing current and steady current, and there is no problem of a frequency band in the measurement.

6.2.1.2 All-Fiber Optical Current Transducer

Usually, an all-fiber optical current transformer (AOCT) uses the Sagnac interference principle to complete the measurement. The input and output optical paths pass through the same fiber and the anti-interference ability is greatly improved. Its model structure is shown in Figure 6.11.

The AOCT utilizes the optical fiber as the sensing material. The optical fiber is then wound around the current wire to be measured from a closed optical circuit. A superluminescent diode (SLD) is used as the light source. The light emitted by the light source is polarized by the fiber polarizer after passing through the coupler to form linearly polarized light. Moreover, the linearly polarized light is injected into the polarization-maintaining fiber at 45° for the x-axis and y-axis transmission, and is then output to the phase modulator for initial phase modulation.

When the two orthogonal modes are linearly polarized, lights pass through the $\lambda/4$ wave plate and are then converted into left-handed and right-handed circularly polarized lights, which are passed to the sensing fiber. Due to the Faraday effect of the magnetic field generated by the current, the two circularly polarized lights are transmitted at different speeds, resulting in a phase difference. After the two beams (of circularly polarized light) are reflected by the end face of the sensing fiber, the polarization modes are interchanged (that is, left-handed light becomes right-handed light, and right-handed light becomes left-handed light). The interaction occurs again, doubling the resultant phase difference. It is restored to a linearly polarized light through the $\lambda/4$ wave plate again. Interference occurs in the fiber polarizer and a coupler is used for detection.

6.2.2 Voltage Measurement with the Optical Technique

6.2.2.1 Principle of the Optical Voltage Transducer

Optical voltage transformers are categorized into three types based on the principles of the electro-optic Pockels effect, the Kerr effect and the inverse piezoelectric effect [21, 25–27]. At present, theoretical research and demonstration applications are focused on the electro-optic Pockels effect. The crystal medium is isotropic without applied voltage,

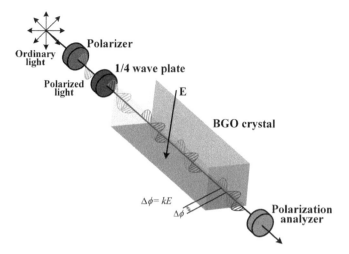

Figure 6.12 The birefringence phenomenon of the Pockels effect.

but it becomes an anisotropic biaxial crystal under the applied voltage. This results in changes in the refractive index and the polarization state of the passing light, bringing in birefringence. Additionally, a beam of light transforms into two linearly polarized lights. The phase difference of the two linearly polarized lights can be expressed as Eqs. (6.3) and (6.4), respectively, with the horizontal electro-optic modulation and longitudinal electro-optic modulation:

$$\delta = \frac{2\pi}{\lambda} n_0^3 \gamma_{41} \frac{l}{d} U = \frac{\pi}{U_{\pi 1}} U \tag{6.3}$$

$$\delta = \frac{2\pi}{\lambda} n_0^3 \gamma_{41} U = \frac{\pi}{U_{\pi 1}} U \tag{6.4}$$

where λ is the wavelength of the light wave; n_0 and γ_{41} are the refractive index and refraction coefficient of the crystal respectively; d and l are the crystal thickness and the through-light length respectively; U is the applied voltage; and $U_{\pi 1}$ and $U_{\pi 2}$ are the half-wave voltages required to create a π phase difference between the two beams, respectively (Figure 6.12).

The polarization interference is converted into the phase difference and the change of output light intensity. The measured voltage can be obtained by photoelectric conversion and the corresponding signal processing.

The electro-optic crystal type of optical voltage transformer makes full use of the Pockels effect to evaluate the voltage measurement. The crystal, under the effect of an applied electric field, is essentially a phase retarder. The main optical devices include a polarizer, a $\lambda/4$ wave plate, an electro-optic crystal and an analyzer. Among them, the polarizer and the analyzer are placed at both ends of the optical path system to form a polarization interference detection system, which is used to measure the phase difference caused by the linear electro-optic effect in the optical path.

The optical voltage transformer based on the Pockels effect can directly measure non-contact high-voltage electric fields without the need for a capacitive voltage division transformer.

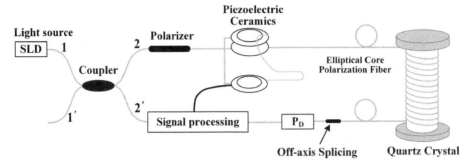

Figure 6.13 Basic structure of an all-fiber type optical voltage transformer. Source: Adapted from Zhi-hong [28].

6.2.2.2 All-Fiber Optical Voltage Transducer

The all-fiber optical voltage transformer (AOVT) is based on the inverse piezoelectric effect of quartz crystal. The measured voltage is applied to the metal electrodes at both ends of the quartz crystal to generate radial strain. Furthermore, the elliptical core dual-mode fiber is stalked and sensed. The phase difference is modulated between the two conduction modes in the fiber using the homodyne phase tracking technique to measure the amount of phase modulation in order to obtain the magnitude and phase of the measured voltage. The structure of the inverse piezoelectric optical voltage transformer is shown in Figure 6.13. The system consists of a sensor head, a light source, a phase tracker, and an interferometer. When an alternating voltage is applied to the sensor head, which is made of quartz crystal, an alternating piezoelectric stress will be generated in a certain direction, and the perimeter of the crystal is modulated. Two modes (LP01 and LP11) propagate in the dual-mode fiber and produce a phase difference as

$$\Delta\varphi = -\pi\frac{Nd_{11}EL_t}{\Delta L_{2\pi}} \tag{6.5}$$

where N is the number of turns of the fiber; E is the electric field strength; L_t is the piezoelectric coefficients; and d_{11} and $\Delta L_{2\pi}$ are, respectively, the piezoelectric coefficient of the crystal and the optical fiber length variation when the crystal generates a 2π phase difference.

By adopting low-coherence interferometry, the phase tracker composed of piezoelectric ceramics is used to realize the modulation phase after selecting the appropriate dual-mode fiber length. Then the voltage magnitude and phase are measured according to the control voltage of the phase tracker.

The transmission and detection of the signal are done in optical fibers to an all-fiber optical voltage transformer where no polarization optical components are required. Therefore, the processing technology is greatly simplified.

6.2.3 Merits and Drawbacks

From the perspective of future development, optical transformers can effectively promote the progress of intelligent equipment and a secondary protection control strategy, in addition to its advantages of a wide dynamic measurement range and immunity to high voltage insulation. Owing to the merits of small size, light weight, simple insulation, etc., it is easy to be integrated with high voltage apparatus like gas insulated switchgear (GIS) and power transformers. It is beneficial to improve the flexibility of equipment installation and effectively save the area of the substation. With the digitized and precise characteristics of optical measurement, the digitization of information collection and the transmission fiber network for the substation are realized, which simplify the wiring and improve the ability to resist electromagnetic interference. Moreover, it effectively supports the rapidity, sensitivity, and reliability and promotes new protection based on transients, sampled values, and the practical application of a power system dynamic observation technology.

As for an all-fiber OVT, the phase sensing head and phase tracker are composed of an elliptical core dual-mode fiber and a piezoelectric crystal, where a significant number of crystal elements is used. At the same time, the wavelength of the transmitted light wave and the length of the multi-mode fiber need to be adjusted reasonably to complete the two-mode transmission and interference. Further practical research and various impact mechanisms are still expected to be done.

Regarding all-fiber OCTs, the Verdet constant of an ordinary silicon fiber is small, and the change of the optical polarization state caused by the inherent birefringence of the fiber tends to drown the Faraday rotation angle. To increase the sensitivity, it is necessary to increase the number of sensing fiber turns. However, this will increase the linear birefringence caused by the intrinsic and bending effects and hence the sensor sensitivity becomes lower than the theoretical value. In addition, the birefringence and Verdet constant of the fiber are also functions of temperature, which further increases the complexity of its practical usage.

6.3 Electric Field Measurement

It is also very necessary to evaluate the electric field and the resultant transient voltage with regard to the lightning and switching overvoltage of power transformers, which is beneficial in an accident analysis and insulation design. Currently, a capacitive voltage divider can be used to detect the transient waveforms in a power transformer with the help of the capacity effect, which acts as the high voltage part in the transformer bushing. As for the low voltage part, a capacitor bank connected with the test tap of the transformer bushing is used [29]. Based on the electric field induced linear birefringence (Pockels effect) in electro-optical crystals, or the converse piezoelectric effect in a cylinder-shaped quartz crystal, a kind of all-fiber (input and output are all fibers), metal-free, and contactless integrated electro-optic field sensors have also been proposed and demonstrated in field applications [30].

Some crystals with the Pockels effect are the key components in an electric field sensor that typically includes lithium niobate ($LiNbO_3$), bismuth titanate $Bi_{12}TiO_{20}$ (BTO), bismuth germanate $Bi_4Ge_3O_{12}$ (BGO), potassium dihydrogen phosphate (KDP), bismuth silicon oxide $Bi_{12}SiO_{20}$ (BSO), etc. The BGO crystals are frequently used due to their stable behavior over a long period of time. The schematic diagram is depicted in Figure 6.14, where PM fiber is short for polarization maintaining optical fiber.

The refractive index of the crystals, electro-optic material in nature, varies as a function of the external electric field and induces a change in the phase of the light beam passing through the crystal. The relationship between the modification in phase and the

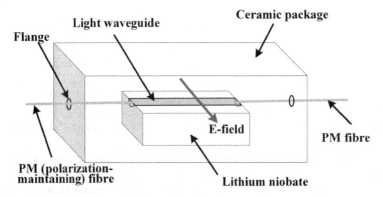

Figure 6.14 Schematic diagram of an integrated optical electric field sensor.

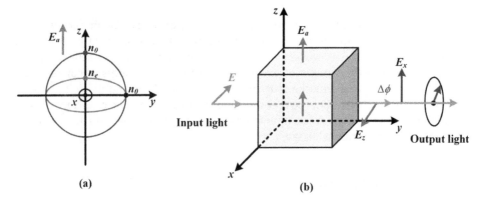

Figure 6.15 (a) Birefringence phenomenon in an isotropic material and (b) a Kerr cell phase modulator.

polarization of the beam can be established to reflect the strength of the electric field. The integrated electric field sensor can then be engaged in the transient voltage under severe high-voltage conditions with the inherent immunity from electromagnetic interference for metering and relaying applications [29, 31, 32].

Due to the robust dielectric performance of oil-pressboard insulation, it has been widely used in high-voltage and ultra-high voltage power transformers, and especially in the newly manufactured converter transformer, where it is extremely important to investigate the electric field distribution measurement, distribution, and allowable strength in the insulation system. Any insert or contact may exert an influence on the original electric field, but the optical non-contact measurement in a liquid dielectric medium is considerably the most ideal solution.

Based on the quadratic electro-optic (QEO) effect, also called the Kerr effect, the applied electric field can be measured by the resultant change in the refractive index of a material. Distinct from the Pockels effect, the induced index change is directly proportional to the square of the electric field instead of varying linearly with it (Figures 6.15–6.18).

More precisely, the phase difference (θ) caused by Kerr effect can be described by Eq. (6.6), in which B denotes the Kerr constant of the liquid dielectric medium

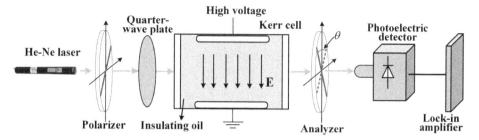

Figure 6.16 Electric field measurement in an insulation oil based on the Kerr effect.

(transformer oil), E represents the field strength, and L is the length of the electric field [33, 34]:

$$\theta = 2\pi B \int_0^L E^2(l)dl \tag{6.6}$$

It has been verified that the electric field distribution of DC, AC, and impulse voltage source is mapped using the Kerr effect-based electric field measurement system [33–37]. The dielectric constant of the insulation medium dominated the AC electric field distribution. However, the insulation medium resistivity plays an important role in the electric filed distribution, which is also influenced by the accumulative behavior and the distribution of the space charges. The electric field distribution under an impulse voltage is usually held equivalent to that of the capacitive field.

Since the analysis and design methodologies for some special transformers or insulation structures remain comparatively inadequate, the electric field distribution of the oil-pressboard insulation system under various voltage waveforms provides vital reference for the design of transformer insulation. It is noteworthy that the measurement system is mainly used as an approach to get an insight into the electric field distribution in the laboratory instead of a mature product.

6.4 Conclusion

In Chapter 6, mechanical aspects like winding deformation and vibration and basic electrical parameters including current, voltage, and electric field with optical approaches are mainly discussed.

Although winding deformation and vibration are closely related, vibration comes from multiple sources such as the magnetic core, winding, cooling fan, etc. FBG and distributed optical sensing can be used to evaluate the winding deformation inside the transformer through the pressure or strain effect, while vibration can be interpreted by interferometric structures and FBG. The demodulation frequency is much lower than a partial discharge and a low cost scheme is possible. Moreover, the installation of multiple vibration probes is extremely essential to locate the defect source or region. The development of deformation and vibration measurement not only solves the theoretical and technical problems of related optical fiber sensing techniques applied to a transformer with technical support for transformer operation and maintenance but also provides new ideas for accurate winding deformation location and fault diagnosis.

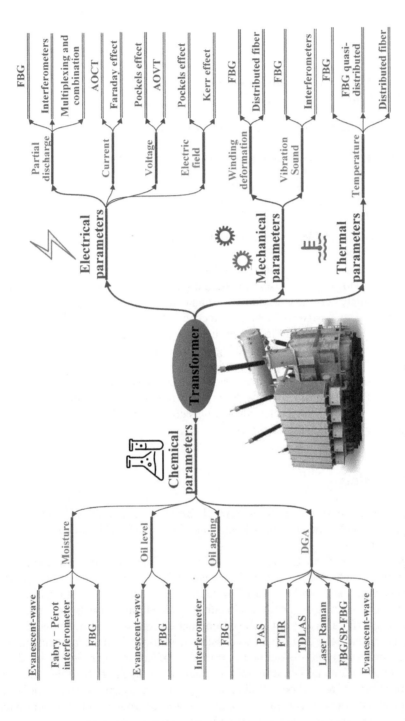

Figure 6.17 Various optical techniques for the measurement in power transformers.

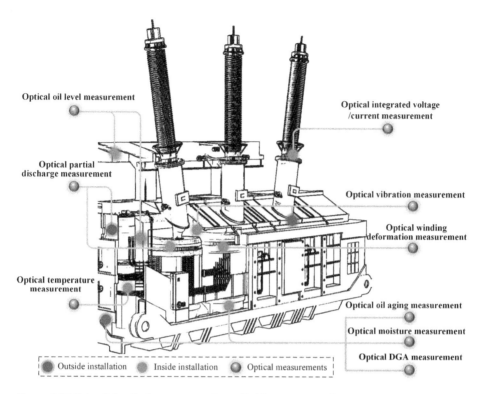

Figure 6.18 Potential optical measurements applied for power transformers.

There are specific current and voltage transducer apparatus in a substation. However, it is believed that in the future it is possible for all-fiber optical current and voltage transducers to be integrated into a power transformer with the advantages of instinct safety and small volume. Particularly, based on the effects of Pockels and Kerr, the electric field strength is measurable with optical techniques that are highly beneficial in providing design fundamentals of complex converter transformers in high voltage direct current (HVDC) scenarios.

6.5 Outlook

Power transformers constitute a precious and critical asset for the power grid, although failures might impose a serious threat or a potential disaster to the system. However, power transformers have to experience thermal, electrical, and mechanical stresses during their lifetime, which leads to a gradual degradation of the paper/oil insulation system. Although partial discharge, overheat, moisture, winding deformation, etc., do harm to an insulation system, any unexpected failures or outages can be avoided with cautious, precise, and comprehensive monitoring. Optical techniques present inspiring solutions and versatile applications in power transformers in this book.

Compared to the existing and conventional methods, the optical solutions are immune to the harsh conditions in a high voltage situation, with flexible installation of inside and outside tanks, light weight materials, and extremely small volume, and so on. Last but not least, optical sensors are compatible with communication systems and have the capacity to carry out remote sensing. A comprehensive application of the optical techniques has been mentioned and illustrated in the chapters, such as FBG, interferometric, scattering, etc. Optical sensing measurement is expected to be a powerful and effective tool to support and even replace conventional techniques. Several challenges and potential directions can be addressed in academic and practical perspectives as shown in the following sections.

6.5.1 Profound and Extensive Interdisciplinary Combinations

Optical fiber does not initially appear to be relevant to power transformers, but it is advantageous to combine the interdisciplinary topics. Although various optical techniques have been tested to serve as sensing approaches to different parameters, they are still in their primary state. Progress in the optical field including optical devices, principles, topologies, etc., is rapidly updated, and sensor detection in power equipment still needs to be followed up substantially. Even the proposed optical technique is able to be combined with other materials, topologies, and mechanical structures, in order to improve the sensitivity, package reliability, etc. As an example, the sensitivity can be enhanced by using pairs of FBGs as the reflectors of an FPI.

In addition, some other subjects and concepts may help through combining advantages and by-passing the different disadvantages in the optical sensors of power transformers. For example, the popular graphene materials, Quantum components, and miniaturized structures can be used to improve the sensing system. The MEMS (micro-electro-machining system) technique is very suitable to cooperate with optic fibers to make sensors more reliable, flexible, and cheaper. In particular, optical fiber type MEMS sensors have further advantages.

6.5.2 Mature Scheme and Low Cost Manufacturing

At present, various optical techniques can be used for one measurand; their merits and drawbacks have been discussed in every section. Sensors do not necessarily need to have different types and principles as they need individual interrogator or detection systems. Limited mature sensing schemes with comprehensive consideration of merits and shortcomings are specifically designed and verified by academic research. The sensing measurement can be developed and manufactured as products. Currently, the market share of optical systems is still relatively low and the price of optical products are always too high.

Optimized optical schemes are the milestones, and economic volume production needs to be organized. To realize the optic-based condition monitoring of power transformers, fiber optic sensor technology has often been driven by the development and subsequent mass production of components that support these industries in turn.

6.5.3 Reliable Measurement and Long-Term Stability

Apart from the cost, another huge barrier in the road to application is the reliability of the optical sensing system, which is also seen as an aspect of immaturity at the present level. Fiber optical sensors have been developed rapidly and have been successfully applied in different apparatus, including power transformers. However, there are still windows for improvement, such as resolution and sensitivity. The effective information is normally poor in industrial substation scenarios, such as the acoustic emission induced by the partial discharge and the concentration of dissolved gases in oil. Highly sensitive measurements are in urgent anticipation. This demand has received significant attention from researchers, which is the reason why many researches have been conducted and highly sensitive prototypes have been developed. Moreover, the environmental ruggedness and field interferences should be emphasized to meet the needs and requirements of sensitivity, which is sometimes very susceptible in real applications.

On the other hand, although the fiber itself is stable and sustainable, its reliability and long-term performance are substantially dependent on the sensing system; that is, optical components are vulnerable to ambient parameters. Therefore, on-going efforts and academic research are necessary to demonstrate the long-term reliability of fibers and improve their presence in the industry.

6.5.4 Pre-factory Installation and Integration into a Monitoring System

The optical sensors interface inputs to power transformers are mentioned extensively in this book, given that it is a critical challenge to the existing structure of transformers. There is no doubt that fiber optics monitoring solutions provide the added value to different types of transformers, but, after all, they are new to traditional power equipment. The power utilities are at risk with the installation of the optical measurement inside the equipment during the in-service period, but pre-factory installation is a good choice with some specific consideration of reserved penetrates or ports during the manufacturing process.

As to the output, the interface with end devices including meters and relays should reconsider the communications protocol and monitoring strategies since the signals of the optical transducers are vastly different from the outputs of conventional instruments. Corresponding standards and guidelines of optical techniques are strongly recommended to be prepared at the current stage.

6.5.5 Rapid Expansion and Development

Current research on optical techniques are mostly focused on large-scale oil-immersed power transformers, but online optical solutions are easily expanded to other power equipment, like converter power transformers, distribution transformers, high voltage reactors, GIS, transmission lines and cables, etc. With similar parameters and requirements, the rapid expansion is promising, and it is beneficial to reduce the cost on mature and reliable schemes and components.

Moreover, research on the possibility and integration of multiple function parameters, like a sensor with the ability to simultaneously measure moisture, vibration, and temperature, is also encouraged and the ports for probe installation turn out to be less in number. As for the management of optical measurements, the portable and movable demodulation system or interrogator for some specific distribution measurement is suggested as inspired by sharing economy. Sharing economy helps to reduce the marginal cost.

Nowadays, increasing newly built HVDC and power electric equipment are put into work in power grids and optical sensors and are also ushered into new development opportunities.

6.5.6 Advanced Algorithms and Novel Diagnosis

Although optical techniques are immune to electromagnetic interference in nature, the measurements are still susceptible to various noises during the complete process, where the data processing and management are very important to achieve high sensitivity and precise decisions, especially for multiple functions. The popular artificial intelligence (AI) algorithms can be the powerful tools to deal with the challenge and diagnose the real-time status of power transformers. Moreover, the distributed edge computing is available to improve response times and save bandwidth to carry out monitoring analysis for power transformers. It is assumed that there will be early warning of internal problems of specific power transformers with the help of multiple optical parameters, such as the occurrence of hotspots, winding deformation, vibration due to stresses arising from short-circuit currents, and the recorded intensity and history of these transient events. In this way, accuracy of the diagnosis is also guaranteed.

In summary, owing to the enormous inherent advantages of optical approaches, the research and development of optical measurements applied in high voltage apparatus have made noticeable progress in recent years, which also provide promising solutions to the future condition-based monitoring of various power equipment. The pros and cons have been weighed from the perspective of electrical, chemical, mechanical, and thermal measurement with accessible optical techniques. Several barriers and challenges are summarized from the future prospects to improve the knowledge of new optical monitoring and to guarantee high reliability and high quality of life management of power transformers. Multi-lateral coordination and cooperation are expected from the manufacturers, maintenance staff, utilities, and academic researchers to move forward this valuable and interesting work.

References

1 Zhao, Z., Yao, C., Li, C., and Islam, S. (2018). Detection of power transformer winding deformation using improved FRA based on binary morphology and extreme point variation. *IEEE Transactions on Industrial Electronics* 65 (4): 3509–3519.

2 Hashemnia, N., Abu-Siada, A., and Islam, S. (2015). Improved power transformer winding fault detection using FRA diagnostics – part 2: radial deformation simulation. *IEEE Transactions on Dielectrics and Electrical Insulation* 22 (1): 564–570.

3 Yong, L., Fan, Y., Fan, Z. et al. (2015). Study on sweep frequency impedance to detect winding deformation within power transformer. *Proceedings of the CSEE* 35 (17): 12.

4 Zhang, H., Yang, B., Xu, W. et al. (2014). Dynamic deformation analysis of power transformer windings in short-circuit fault by FEM. *IEEE Transactions on Applied Superconductivity* 24 (3): 1–4.

5 Secue, J.R. and Mombello, E. (2008). Sweep frequency response analysis (SFRA) for the assessment of winding displacements and deformation in power transformers. *Electric Power Systems Research* 78 (6, 1128): 1119.

6 Bagheri, M., Naderi, M.S., Blackburn, T., and Phung, T. (2013). Frequency response analysis and short-circuit impedance measurement in detection of winding deformation within power transformers. *IEEE Electrical Insulation Magazine* 29 (3): 33–40.

7 de Melo, A.G., Benetti, D., de Lacerda, L.A. et al. (2019). Static and dynamic evaluation of a winding deformation FBG sensor for power transformer applications. *Sensors* 19 (22): 4877.

8 Liu, Y., Yin, J., Fan, X., and Wang, B. (2019). Distributed temperature detection of transformer windings with externally applied distributed optical fiber. *Applied Optics* 58 (29): 7962–7969.

9 Liu, Y., Tian, Y., Fan, X. et al. (2018). A feasibility study of transformer winding temperature and strain detection based on distributed optical fibre sensors. *Sensors* 18 (11): 3932.

10 T. Yuan, "Research on transformer winding deformation detection method based on distributed optical fiber sensing," PhD Thesis, North China Electric Power University, North China Electric Power University, Beijing, China, 2019.

11 Garcia, B., Burgos, J.C., and Alonso, A.M. (2006). Transformer tank vibration modeling as a method of detecting winding deformations-part II: experimental verification. *IEEE Transactions on Power Delivery* 21 (1): 164–169.

12 Zhou, H., Hong, K., Huang, H., and Zhou, J. (2016). Transformer winding fault detection by vibration analysis methods. *Applied Acoustics* 114: 136–146.

13 Shengchang, J., Fan, Z., Yuhang, S. et al. (2020). Review on vibration-based mechanical condition monitoring in power transformers. *High Voltage Engineering* 46 (1): 16.

14 Shengchang, J., Yuhang, S., Fan, Z., and Weifeng, L. (2019). Review on vibration and noise of power transformer and its control measures. *High Voltage Apparatus* 55 (11): 17.

15 Min, L., Li, S., Zhang, X. et al. (2018). The research of vibration monitoring system for transformer based on optical fiber sensing. In: *2018 IEEE 3rd Optoelectronics Global Conference (OGC)*, 126–129. IEEE.

16 Kung, P., Idsinga, R., Fu, J.B. et al. (2016). Online detection of windings distortion in power transformers by direct vibration measurement using a thin fiber optics sensor. In: *2016 IEEE Electrical Insulation Conference (EIC)*, 576–578. IEEE.

17 Gangopadhyay, T.K. (2004). Prospects for fibre Bragg gratings and Fabry-Perot interferometers in fibre-optic vibration sensing. *Sensors and Actuators A: Physical* 113 (1): 20–38.

18 Garcia-Souto, J.A. and Lamela-Rivera, H. (2006). High resolution (< 1nm) interferometric fiber-optic sensor of vibrations in high-power transformers. *Optics Express* 14 (21): 9679–9686.

19 Rivera, H.L., Garcia-Souto, J.A., and Sanz, J. (2000). Measurements of mechanical vibrations at magnetic cores of power transformers with fiber-optic interferometric intrinsic sensor. *IEEE Journal of Selected Topics in Quantum Electronics* 6 (5): 788–797.

20 Raghavan, A., Kiesel, P., Teepe, M. et al. (2020). Low-cost embedded optical sensing systems for distribution transformer monitoring. *IEEE Transactions on Power Delivery*: 1.

21 Katsukawa, H., Ishikawa, H., Okajima, H., and Cease, T.W. (1996). Development of an optical current transducer with a bulk type Faraday sensor for metering. *IEEE Transactions on Power Delivery* 11 (2): 702–707.

22 Xu, Q., Xie, N., Wang, D., and Huang, Y. (2018). A linear optical current transducer based on Newton's ring sub-wavelength grating. *IEEE Sensors Journal* 18 (17): 7041–7046.

23 Ulmer, E.A. (1990). A high-accuracy optical current transducer for electric power systems. *IEEE Transactions on Power Delivery* 5 (2): 892–898.

24 Silva, R.M., Martins, H., Nascimento, I. et al. (2012). Optical current sensors for high power systems: a review. *Applied Sciences* 2 (3): 602–628.

25 Delle Femine, A., Gallo, D., Landi, C., and Luiso, M. (2007). Broadband voltage transducer with optically insulated output for power quality analyses. In: *2007 IEEE Instrumentation & Measurement Technology Conference IMTC 2007*, 1–6. IEEE.

26 Chen, J., Xie, L., and Song, J. (2004). Recent research on optical voltage transducer [J]. *China Measurement Technology* 3: 45–50.

27 Monsef, H. and Ghomian, T. (2006). Modified quadrature method for accurate voltage measurement in optical voltage transducer. *IEE Proceedings-Generation, Transmission and Distribution* 153 (5): 524–530.

28 Zhi-hong, X. (2014). Study and comment of the optical transformers in power system. *Power System Protection and Control* 42 (12): 7.

29 Wang, H., Zhuang, C., Zeng, R. et al. (2019). Transient voltage measurements for overhead transmission lines and substations by metal-free and contactless integrated electro-optic field sensors. *IEEE Transactions on Industrial Electronics* 66 (1): 571–579.

30 Xie, S., Zhang, Y., Yang, H. et al. (2019). Application of integrated optical electric-field sensor on the measurements of transient voltages in AC high-voltage power grids. *Applied Sciences* 9 (9): 1951.

31 Yang, Q., Sun, S., Han, R. et al. (2015). Intense transient electric field sensor based on the electro-optic effect of LiNbO3. *AIP Advances* 5 (10): 107130.

32 N'cho, J.S. and Fofana, I. (2020). Review of Fiber optic diagnostic techniques for power transformers. *Energies* 13 (7): 1789.

33 Qi, B., Gao, C., Zhao, X. et al. (2016). Interface charge polarity effect based analysis model for electric field in oil-pressboard insulation under DC voltage. *IEEE Transactions on Dielectrics and Electrical Insulation* 23 (5): 2704–2711.

34 Qi, B., Zhao, X., Li, C., and Wu, H. (2015). Transient electric field characteristics in oil-pressboard composite insulation under voltage polarity reversal. *IEEE Transactions on Dielectrics and Electrical Insulation* 22 (4): 2148–2155.

35 Qi, B., Zhao, X., and Li, C. (2016). Methods to reduce errors for DC electric field measurement in oil-pressboard insulation based on Kerr-effect. *IEEE Transactions on Dielectrics and Electrical Insulation* 23 (3): 1675–1682.

36 Qi, B., Zhao, X., Li, C., and Wu, H. (2016). Electric field distribution in oil-pressboard insulation under AC-DC combined voltages. *IEEE Transactions on Dielectrics and Electrical Insulation* 23 (4): 1935–1941.

37 Qi, B., Zhao, X., Zhang, S. et al. (2017). Measurement of the electric field strength in transformer oil under impulse voltage. *IEEE Transactions on Dielectrics and Electrical Insulation* 24 (2): 1256–1262.

Index

Optical Sensing in Power Transformers, First Edition. Jun Jiang and Guoming Ma.
© 2021 John Wiley & Sons Ltd. Published 2021 by John Wiley & Sons Ltd.